T0201588

Tectonic, Climatic, and Cryospheric Evolution of the Antarctic Peninsula

John B. Anderson and Julia S. Wellner
Editors

American Geophysical Union

Library of Congress Cataloging-in-Publication Data

Tectonic, climatic, and cryospheric evolution of the Antarctic Peninsula / John B. Anderson and Julia S. Wellner, editors.
    p. cm. — (Special publication ; 63)
  Includes bibliographical references and index.
  ISBN 978-0-87590-734-5
  1. Antarctica. 2. Geology—Antarctica. 3. Geology, Stratigraphic—Cenozoic. 4. Morphotectonics—Antarctica. 5. Climatic changes—Antarctica. I. Anderson, John B., 1944- II. Wellner, Julia S.
  QE350.T42 2011
  559.89—dc23
                            2011040153

  ISBN: 978-0-87590-734-5

  Book doi: 10.1029/SP063

CONTENTS

PREFACE

SHALDRIL I and SHALDRIL II presented some formidable logistical challenges. However, thanks to the hard work of many people, we were able to sample strata from key time intervals and obtain a record of climate change and associated changes in plants living on and near the continent. This volume contains detailed results from analyses of the drill core, but it also contains papers that present the seismic stratigraphic approach to drilling and papers that focus on the tectonic evolution of the Antarctic Peninsula, which strongly influenced climate change. It represents a synthesis of research by 20 scientists, and we are grateful to all of the authors for their contributions to this volume. We also thank those individuals who provided constructive and timely reviews of the papers. Finally, we wish to thank those people who helped with the logistics, participated on the cruises, and assisted with the research and publication of results. There are too many to name all, but we greatly appreciate their hard work and positive attitudes during the best and the worst of times. We would especially like to thank those people that did so much to make these first drilling legs happen: Leon Holloway, Jim Holik, Ashley Lowe Ager, and Andy Frazer.

John B. Anderson
Rice University

Julia S. Wellner
University of Houston

Tectonic, Climatic, and Cryospheric Evolution of the Antarctic Peninsula
Special Publication 063
Copyright 2011 by the American Geophysical Union.
10.1029/2011SP001119

Introduction

John B. Anderson

Department of Earth Science, Rice University, Houston, Texas, USA

Julia S. Wellner

Department of Earth and Atmospheric Sciences, University of Houston, Houston, Texas, USA

Antarctica's climate and glacial history remain shrouded due largely to a paucity of outcrops and drill cores; in particular, the Neogene record is fragmentary. Most outcrops of this age have been studied at some level of detail, so the greatest opportunity for expanding our knowledge of this time interval is through acquisition of drill cores. Drill cores from the deep-sea floor around Antarctica have provided a proxy record of climate change and ice sheet evolution but do not allow us to address questions concerning regional variability in ice sheet development or the response of organisms living on the continent to climate change. The continental shelf contains a rich and virtually unsampled stratigraphic record, but sea ice and icebergs limit access to these areas by conventional drill ships. This has prompted efforts to drill the continental shelf using unconventional methods, including drilling from the sea ice (ANDRILL) and from ice-breaking research vessels (SHALDRIL).

An understanding of the Cenozoic history of the Antarctic continent and its ice sheets has been hindered by a scarcity of outcrops that are not covered by thick ice, deep water, stiff till, or some combination of those three. Great advances have been made in recent decades in studies of the current Antarctic Ice Sheet both on the ice and from space and the Holocene records around the Antarctic margin, as well the oceans and biota that surround the continent. Despite the many advances, the older history of the continent has remained a relative mystery because of the difficulty in accessing the rocks that hold the record. The few records that are obtained are key inputs to modeling studies of the Antarctic Ice Sheet history [e.g., *Pollard and DeConto*, 2009]. The lack of other long-term records of ice and climate fluctuations around the Antarctic margin hinders the development and accuracy of such ice sheet models and our ability to look toward the future behavior of the ice sheet.

Tectonic, Climatic, and Cryospheric Evolution of the Antarctic Peninsula
Special Publication 063
Copyright 2011 by the American Geophysical Union.
10.1029/2011SP001132

1

Around the Antarctic continent are a series of seaward dipping strata, many of which have been deeply eroded during advances of the Antarctic ice sheets. Pre-Pleistocene strata lie just beneath the seafloor, and their distribution can be mapped by high-resolution seismic methods [cf. *Anderson*, 1999]. The concept behind SHALDRIL (shallow drilling) is to drill through surficial glacial deposits and sample older strata where they come close to the seafloor. SHALDRIL was never meant for obtaining long cores but, rather, to collect a series of relatively short cores (~100 m) that can be pieced together using seismic records and detailed chronostratigraphy and combined with seismic stratigraphic information to reconstruct climatic and cryospheric history.

SHALDRIL began in earnest in 1994 with a workshop sponsored by the U.S. National Science Foundation (NSF) and held at Rice University. At that time, the prevailing conclusion was that the technology was not yet ready for putting a drilling system onto a U.S. Antarctic Program (USAP) icebreaker: certainly, there were systems that could have been employed but not with a reasonable chance of success under the harsh conditions expected. Any drilling on the Antarctic shelf had to contend with pebbly glacial tills, which are notoriously difficult to drill anywhere, freezing conditions, drifting icebergs, and substantial sea ice cover, as well as the fact that any icebreaker brought to the task had not been designed for drilling. A committee was formed to monitor the technology, and in 2001 a proposal was sent to the NSF for a demonstration drilling cruise. This proposal became SHALDRIL I and SHALDRIL II.

The research vessel icebreaker *Nathaniel B. Palmer* (NBP), the primary scientific icebreaker used by USAP at the time, was chosen as the vessel to be used for the drilling legs. The vessel was modified by the installation of a moon pool, through which drilling operations could take place and, later, the introduction of additional ballast in order to accommodate the weight of the drill rig. These were permanent changes to the vessel, and the moon pool was subsequently used by a variety of other scientific programs. Seacore (later Fugro Seacore) of Cornwall, United Kingdom, was hired to perform drilling operations. A drilling rig was custom-designed to fit on the NBP, and this rig was leased for the duration of two drilling seasons, cruises of about 5 weeks duration each during the austral falls of 2005 and 2006.

SHALDRIL I had its first drilling target in Maxwell Bay in the South Shetland Islands and quickly recovered a long (108 m) core in soft Holocene sediments with high recovery percentages. Unfortunately, this quick success would prove to be one of the only scientific successes of the first SHALDRIL season. The drilling tools employed during the first season of SHALDRIL made very slow progress when drilling through the glacial till, which, in varying thicknesses, covered all of the older targets in the area east of James Ross Island in the northwestern Weddell Sea. Drifting icebergs made it impossible to stay on station long enough for the tools to penetrate the till, and no pre-Pleistocene material was recovered during SHALDRIL I. Nonetheless, two additional Holocene cores were recovered: one each from Herbert Sound and Lapeyrère Bay. The three Holocene cores obtained during 2005 plus the Firth of Tay core collected in 2006 together make up the most detailed record of Holocene glacial and climate history around the Antarctic Peninsula; they are documented in several papers [e.g., *Michalchuk et al.*, 2009; *Milliken et al.*, 2009] and are parts of other ongoing studies. SHALDRIL I had technical successes also, reported in the cruise report (http://www.arf.fsu.edu/projects/shaldril.php) and short articles in *Scientific Drilling* and other reports [e.g., *Wellner et al.*, 2005], as well as a great success obtaining Holocene records. The knowledge base built during SHALDRIL I allowed modifications to the drilling equipment and to the general approach employed while drilling and thus set the stage for the 2006 drilling leg.

In the 12 months between SHALDRIL I and SHALDRIL II, substantial changes were made to the drill bits and downhole sampling tools used by the program. The modifications improved the ability of the drill to penetrate through glacial sediment and to obtain samples below it. The sea ice in the Weddell Sea during early 2006, though, proved to be thicker and much more extensive than normal. Even worse, the sea ice and the abundant icebergs were moving rapidly. One core, Site 3, sampled Eocene sediments off of James Ross Island at the targeted site. This and other cores were made possible not only by the new equipment and quick actions of the drillers but also by flexibility to move sites based on ice conditions. Further reductions in the time needed at any one site were made by moving the ship with the drill pipe hanging below; drill pipe was assembled to a length just several meters above the seafloor as the vessel continued to maneuver around ice, thus eliminating the time needed to trip the pipe from the window of ice conditions at any one site. Despite such flexibility, the ice conditions around the primary targets in the James Ross Basin were so severe that plans for the remaining drill targets had to be aborted. However, backup sites along the southern margin of the Joinville Plateau yielded Oligocene, Miocene, Pliocene, and Pleistocene strata. Selection of these new targets was made possible by collecting and interpreting additional seismic data during the cruise. The cores recovered during this second leg have provided a record of climatic change and cryospheric evolution that spans the latest Eocene through the Pleistocene, although there are significant gaps between each core.

This volume represents the summary of the scientific results of the SHALDRIL II program as well as contributions from other authors studying the tectonic history of the region, particularly as it relates to the establishment of ocean gateways and mountain building. These results, along with those from ANDRILL and the few Integrated Ocean Drilling Program legs that have made it to the Antarctic shelf, are bringing the stratigraphic history of the Antarctic to light. It is hoped that these results are just the start, and now that the efficacy of the SHALDRIL approach has been demonstrated, there will be more drill cores recovered in the near future, including a more complete record from the James Ross Basin, of which SHALDRIL II has provided just a tantalizing glimpse.

REFERENCES

Anderson, J. B. (1999), *Antarctic Marine Geology*, 289 pp., Cambridge Univ. Press, Cambridge, U. K.

Michalchuk, B., J. B. Anderson, J. S. Wellner, P. L. Manley, S. Bohaty, and W. Majewski (2009), Holocene climate and glacial history of the northeastern Antarctic Peninsula: The marine sedimentary record from a long SHALDRIL core, *Quat. Sci. Rev.*, *28*, 3049–3065, doi:10.1016/j.quascirev.2009.08.012.

Milliken, K. T., J. B. Anderson, J. S. Wellner, S. Bohaty, and P. Manley (2009), High-resolution Holocene climate record from Maxwell Bay, South Shetland Islands, Antarctica, *Geol. Soc. Am. Bull.*, *121*(11–12), 1711–1725, doi:10.1130/B26478.1.

Pollard, D., and R. M. DeConto (2009), Modeling West Antarctic Ice Sheet growth and collapse through the past five million years, *Nature*, *458*, 329–332.

Wellner, J. S., J. B. Anderson, and S. W. Wise (2005), The inaugural SHALDRIL expedition to the Weddell Sea, Antarctica, *Sci. Drill.*, *1*, 40–43.

J. B. Anderson, Department of Earth Science, Rice University, Houston, TX 77005, USA. (johna@rice.edu)

J. S. Wellner, Department of Earth and Atmospheric Sciences, University of Houston, Houston, TX 77204, USA.

A Different Look at Gateways:
Drake Passage and Australia/Antarctica

Lawrence A. Lawver, Lisa M. Gahagan, and Ian W. D. Dalziel

Institute for Geophysics, University of Texas at Austin, Austin, Texas, USA

The time of the opening of Drake Passage between South America and the
Antarctic Peninsula is problematic. Mammals were able to migrate between
South America and Antarctica until sometime in the early Eocene. Various
continental fragments may have formed an effective barrier to substantial
deepwater circulation through Drake Passage until at least 28 Ma. Alternatively, a medium-depth to deep water passage may have existed through
Powell Basin to the south of the present Drake Passage as early as 33 Ma,
but it is difficult to constrain the time of opening of Powell Basin. Simple
opening of a shallow seaway between southern South America and the
Antarctic Peninsula does not produce a vigorous Antarctic Circumpolar
Current (ACC). Other gateways must be open to medium-depth to deepwater
circulation such as one between the South Tasman Rise and East Antarctica.
Even mid-ocean plateaus may play a role in the ultimate development of a
circum-Antarctic current. The most probable southern ocean feature that may
have affected global circulation was the opening of a deep seaway between
the Kerguelen Plateau and Broken Ridge at about the Eocene-Oligocene
boundary. While a complete deepwater (>2000 m) circuit was certainly
developed by the end of the early Oligocene, it may have been the closure
of a major deep seaway north of Australia in the middle Miocene that finally
produced the environment for the development of a vigorous ACC.

1. INTRODUCTION

The opening of southern ocean gateways has long been considered a significant factor in not only the initiation of the Antarctic Circumpolar Current (ACC) but also in the development of the Cenozoic East Antarctic Ice Sheet [*Kennett*, 1977]. The Eocene-Oligocene boundary step in the ∂O^{18} anomaly history [*Zachos et al.*, 2001, 2008] at 33.7 Ma (timescale from *Berggren et al.* [1995]) is taken as a proxy to represent the onset of Antarctic bottom water formation at temperatures close to freezing [*Kennett and Shackleton*, 1976] and, in turn, the final opening of

Tectonic, Climatic, and Cryospheric Evolution of the Antarctic Peninsula
Special Publication 063
Copyright 2011 by the American Geophysical Union.
10.1029/2010SP001017

Figure 1. Polar stereographic location map showing the Antarctic Circumpolar Current (ACC) derived from the work of *Sandwell and Zhang* [1989] as black arrows. Deep Sea Drilling Project and Ocean Drilling Project sites are shown as numbers. The present-day plate boundary is shown as a thin gray line, while large igneous provinces are shown in a dark gray. Magnetic anomaly picks are shown as crosses, while magnetic isochrons are shown as lines parallel or subparallel to the plate boundaries. Lines orthogonal to the plate boundaries are fracture zone lineations picked from bathymetry derived from satellite altimetry data [*Smith and Sandwell*, 1997]. Abbreviations are BR, Broken Ridge; DP, Drake Passage region; KP, Kerguelen Plateau; PB, Prdyz Bay; LG, Lambert Graben; STR, South Tasman Rise; TAS, Tasmania.

a circum-Antarctic seaway that isolated Antarctica. *Lyle et al.* [2007] support a late Oligocene to early Miocene initiation of an ACC based on sedimentation in the South Pacific backed by grain size evidence from the Tasman Gateway [*Pfuhl and McCave*, 2005]. Based on neodymium isotope ratios at Agulhas Ridge, *Scher and Martin* [2004] support an initial opening of Drake Passage as early as middle Eocene, although *Livermore et al.* [2005] suggest only a possible shallow seaway development at Drake Passage perhaps as early as middle Eocene with a deep seaway only developing between 34 to 30 Ma coincident with the increase in the high latitude ∂O^{18} values at the Eocene-Oligocene boundary. The ACC shown in Figure 1 is presently the largest ocean current with an eastward flow rate through the Drake Passage region of 136.7 ± 7.8 Sv based on the baroclinic transport relative to the deepest common level [*Cunningham et al.*, 2003]. In the region of the Scotia Sea shown in Figure 2, they found that the ACC transport is principally carried in two jets, the Subantarctic Front (SAF), which accounts for 53 ± 10 Sv, and the Polar Front (PF), which accounts for 57.5 ± 5.7 Sv. Southward of the main ACC, they calculated that the Southern Antarctic Circumpolar Current Front (SACCF) transports 9.3 ± 2.4 Sv.

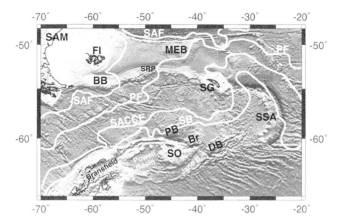

Figure 2. The ACC fronts for the Scotia Sea region, digitized from Figure 1 of the work of *Naveira-Garabato et al.* [2002], superimposed on the bathymetry of the region taken from the most recent online version (August, 2010) of the work of *Sandwell and Smith* [1997]. Fronts are labeled in white: PF, Polar Front; SACCF, Southern ACC Front; SAF, Subantarctic Front; SB, Southern Boundary of the ACC. Geographical features are labeled in black: BB, Burdwood Bank; Br, Bruce Bank; DB, Discovery Bank; FI, Falkland Islands; MEB, Maurice Ewing Bank; PB, Pirie Bank; SAM, South America; SG, South Georgia Island; SO, South Orkney block; SRP, Shag Rocks Passage; SSA, South Sandwich Arc.

The locations of the fronts that define the ACC in the Scotia Sea region (Figure 2) are taken from the work of *Naveira-Garabato et al.* [2002], who have the SAF curling around the eastern end of Burdwood Bank before heading north across the eastern end of the Falkland Plateau and show the PF exiting the central Scotia Sea at a gap in the north Scotia Ridge west of Aurora Bank at about 48°W (labeled the Shag Rocks Passage). Consequently, the majority of transport of the ACC exits across the north Scotia Ridge to the west of 48°W, well west of South Georgia Island. The SACCF curls around the eastern tip of South Georgia to flow westward through the Northeast Georgia Passage of *Naveira-Garabato et al.* [2002]. The remainder of the ACC current flow is bounded to the south and east by the Southern Boundary of the ACC and does not flow directly eastward across the South Sandwich Arc (SSA) as might be expected; rather, it exits the Scotia Sea at the deepest point along the northern margin of the east Scotia Sea at about 30°E. It is clear that substantial ACC transport is dependent on medium-depth to deepwater passageways, seemingly those at least 2700 m deep. An analog to the ACC may be the interocean exchange of thermocline water [*Gordon*, 1986], which runs into a choke point in the Lombok Strait of the Sunda Arc where only 1.7 Sv out of the 7 to 18.6 Sv Indonesian Throughflow [*Gordon and Fine*, 1996; *Gordon et al.*, 2003] passes through the strait which has a sill depth of ~300 m. The remainder of the throughflow is diverted over 1000 km to the east to enter the Indian Ocean via the Timor Trough where a sill depth of 1300 to 1500 m is found. Consequently, the impact of the opening of Drake Passage to deep water flow is critical to understanding its impact on Cenozoic climate. While a shallow Drake Passage may have opened during the Eocene, its impact on the Cenozoic climate may have only become significant when a deepwater passage finally opened.

As shown in the ∂O^{18} compilation of *Zachos et al.* [2008], the world's oceans began to cool after the early Eocene Climatic Optimum (EECO) (equal to ~53 to 51 Ma), well before the Eocene-Oligocene boundary. It is important to determine both when a Cenozoic land barrier first disappeared in the Drake Passage region and when a deep seaway finally developed. With respect to the development of a seaway or conversely, the elimination of a "land bridge" between South America and the Antarctic Peninsula, the best indication may be when mammals were no longer able to disperse between the two continents. *Woodburne and Zinsmeister* [1984] suggested that the origin of the Seymour Island polydolopid mammals based on the endemism and specialization they observed in the two Antarctic genera of Polydolopidae, *Antarctodolops* and *Eurydolops*, as opposed to the Patagonian members of this extinct family occurred roughly 10 million years prior to the fossils they found in TELM5 bed. That bed has been recently redated to be 48 to 51 Ma by *Ivany et al.* [2009] so the age from the work of *Woodburne and Zinsmeister* [1984] would be middle to late Paleocene, and they concluded that the vicariant isolation of the land mammal fauna of the Antarctic Peninsula took place prior to the early Eocene. *Reguero and Marenssi* [2010] inferred that the last mammal dispersal between South America and the Antarctic Peninsula had to have been at the end of Paleocene or during the earliest Eocene based on consideration of marsupials and other mammals. While mammals survived on the Antarctic Peninsula until the very end of the Eocene, ~34.2 Ma [*Bond et al.*, 2006], the 12-fossil mammal taxa found there represent a bimodal size distribution unlike the similar Eocene fauna from Patagonia, suggesting isolation. In addition, the sudamericid *Sudamerica ameghinoi* became extinct in South America by the late Paleocene but survived in Antarctica until the middle Eocene [*Goin et al.*, 2006]. *Reguero and Marenssi* [2010] suggest that the last South American mammals dispersed to Antarctica during the onset of a late Paleocene to early Eocene thermal warming (58.5 to 56.5 Ma) during a time of major regressive events recorded either on the Antarctic Peninsula or southernmost Patagonia. The Antarctic mammal populations became isolated, with the survivors remaining as endemic species. This timing fits with the known migration of marsupials between South America and Australia with the first Australian marsupials, *Djarthia murgonensis*, found in the early Eocene Tingamarra fauna in southeastern Queensland [*Godthelp et al.*, 1999]. They are dated with a minimum age of 54.6 Ma and presumably reached Australia from South America via Antarctica [*Goin et al.*, 2007]. There is no evidence that Australian mammals were able to return to South America [*Nilsson et al.*, 2010] after this time, so initiation of an early Eocene shallow seaway between South America and Antarctica is consistent with an earlier, one-way dispersal of the marsupials to Australia and the probable Late Cretaceous movement of monotremes from Australia to South America [*Pascual et al.*, 2002].

The opening of Drake Passage as an Eocene gateway has been suggested on plate tectonic grounds by both *Livermore et al.* [2005] and by *Eagles et al.* [2006] and on neodymium isotope ratios at Agulhas Ridge [*Scher and Martin*, 2004]. *Livermore et al.* [2005] believe a major change in motion between South America and Antarctica at about 50 Ma led to an early seaway, while *Eagles et al.* [2006] describe two small basins in the southern part of the Scotia Sea, the Protector and Dove basins, which they believe may have opened in the middle or late Eocene and may have been the first seaway to develop between South America and Antarctica. Based on the fact that early Cenozoic Pacific ε_{Nd} values (^{143}Nd/^{144}Nd ratios) are more radiogenic (i.e., ε_{Nd} values -3 to -5) [*Ling et al.*, 1997] than Atlantic values (ε_{Nd} ~-9) [*Thomas et al.*, 2003], there was a steep Pacific-Atlantic gradient [*Goldstein and Hemming*, 2003] at that time. *Scher and Martin* [2004] found a significant increase (from ~-9 to ~-6.4) in the Atlantic values beginning about 41 Ma. They used both ferromanganese (Fe-Mn) crusts and fossil fish teeth to determine ε_{Nd} values for Eocene bottom water and interpret the increase to represent the first influx of Pacific water into the Atlantic at Agulhas Ridge. An even earlier seaway between East and West Antarctica, one that connected Pacific and Atlantic waters, has been suggested by *Casadio et al.* [2010] to be as old as early Cretaceous. *Dalziel and Lawver* [2001] show a Ross Sea/Weddell Sea connection to be as old as Late Cretaceous, formed after the extension in the Ross Sea region ended about 90 Ma. *Ghiglione et al.* [2009] present evidence for the presence of a latest Paleocene-early Eocene extensional basin (i.e., lateral rift) in Tierra del Fuego with a postrift unconformity that indicates extensional faulting ending ~49 Ma. They suggested this basin as a possible early seaway that crossed southern South America. Such an age fits with the increase in South America-Antarctica separation rate found by *Livermore et al.* [2005] and would be immediately prior to the proposed rifting found in the Protector and Dove basins by *Eagles et al.* [2006].

Since there is now reasonable evidence for at least a shallow seaway between southern South America and the Antarctic Peninsula as early as ~50 Ma, based on mammal migration and the isolation of the endemic mammals found in the La Meseta Formation on Seymour Island [*Reguero and Marenssi*, 2010], the timing of the development of a seaway between the South Tasman Rise (STR) and East Antarctica may be the critical factor in the thermal isolation of Antarctica and the eventual development of the ACC as *Kennett* [1977] suggested. Reconstructions of major plate motions indicate that the STR cleared the Oates Coast margin (158°E) of East Antarctica sometime between 34 and 32 Ma [*Lawver and Gahagan*, 2003; *Cande and Stock*, 2004a, 2004b] although *Brown et al.* [2006] show the Tasman Gateway still closed at 32 Ma. The timing suggested by *Lawver and Gahagan* [2003] was based on the assumption that the wide continental shelf off the Wilkes Subglacial Basin of East Antarctica between 145°E and 158°E was in place prior to the development of the East Antarctic Ice Sheet. If, in fact, the outer margin as defined by the Antarctic Margin Gravity High (AMGH) is the product of Cenozoic glaciation on East Antarctica, then opening of a seaway between the STR and East Antarctica may have occurred prior to the Eocene-Oligocene boundary, perhaps as early as 40 Ma. Such an age is in good agreement with the time of opening cited by *Exon et al.* [1995]. In addition to the possibility of a deep water passage south of the STR, there is also a submarine trough, the South Tasman Saddle [*Exon et al.*, 1995; *Exon and Crawford*, 1997, p. 539] between Tasmania and the STR. The present South Tasman Saddle is deepest in a narrow 3000 m deep trough, but the present-day 2000 m contour between Tasmania and the STR defines a seaway almost 100 km wide. At the 3000 m isobath, the South Tasman Saddle is only a few kilometers wide and may have always been very narrow at its deepest part. The South Tasman Saddle may be the result of seafloor spreading or continental extension in the late Cretaceous and would have allowed significant water transport even if it was shallower in the past. The present-day ACC shown in Figure 1 is diverted south and around the Campbell Plateau [*Sandwell and Zhang*, 1989; *Neil et al.*, 2004] even though the plateau is between 700 and 1000 m deep just as it is diverted north around the SSA as shown in Figure 2. A relatively narrow, 1000 m deep trough between Tasmania and the STR might not have allowed major water circulation or the initiation of an early ACC but a 100 km wide, 2000 m deep trough would have.

2. METHODOLOGY

The plate reconstructions used to constrain the opening times of various seaways are derived from a global database, which consists of marine magnetic anomalies tied to the *Gee and Kent* [2007] timescale, paleomagnetic poles, seafloor age dates based on drilling results, and fracture zone and transform fault lineations [*Gahagan et al.*, 1988] picked from ship track and satellite altimetry data [*Sandwell and Smith*, 1997; *Smith and Sandwell*, 1997]. We define continental block outlines based on our own digitization of the satellite altimetry data. Off Namibia, there is a close correlation between the steep gradient in the satellite altimetry data picked by *Lawver et al.* [1998] and the ocean-continent boundary deduced from seismic refraction and reflection data [*Gladczenko et al.*, 1997; *Bauer et al.*, 2000]. In other regions, there may be some stretched continental crust oceanward of the steep gradient that we picked, but for reconstruction purposes, we assume the crust to be predominantly continental landward of the boundary and oceanic, seaward of the line.

The same steep gradient in the satellite altimetry data was used by *Royer and Sandwell* [1989] to determine a limit to the continental shelves in the eastern Indian Ocean. They referred to this

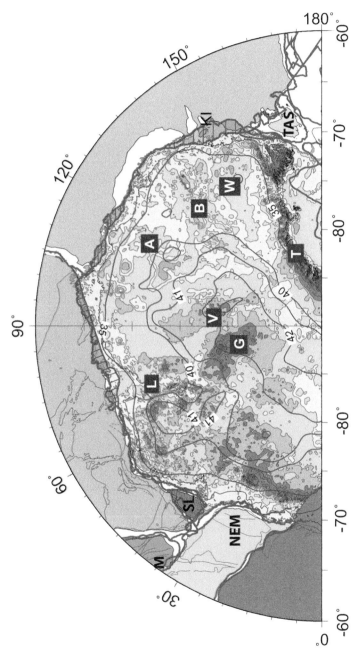

Figure 3. The BEDMAP sub-ice topography of *Lythe et al.* [2000] is shown in the tight-fit reconstruction of Gondwana [*Lawver et al.*, 1998]. The CSBs are shown as thick red lines. Note the excellent match between East Antarctica (EANT) and Australia (AUS) between 120°E and 135°E and the large overlap shown hatched between 135°E and 158°E. Numbered contours on EANT are crustal thickness contours in km (based on the work of *Ritzwoller et al.* [2001]). Abbreviations are A, Aurora Subglacial Basin; B, Belgica Subglacial Highlands; G, Gamburtsev Subglacial Highlands; GAB, Great Australian Bight; KI, Kangaroo Island; MAD, Madagascar; NEM, Northeast Mozambique; SL, Sri Lanka; T, Transantarctic Mountains; V, Vostok Subglacial Highlands; W, Wilkes Land Subglacial Basin.

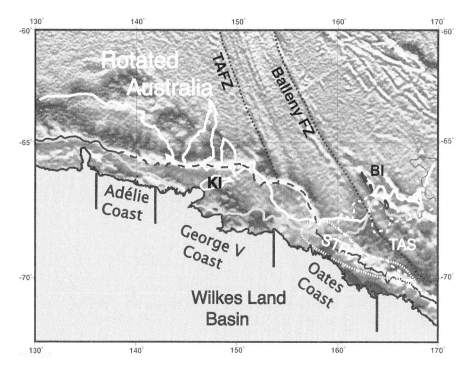

Figure 4. Free-air gravity anomalies derived from satellite altimetry data [*Sandwell and Smith*, 1997] shown. Antarctica is shown in gray. The CSB for Antarctica is shown as a black line or a dashed black line. The rotated Australian coastline is shown as a thicker white line, while the Australian CSB is shown as the thick gray line. KI is shown inboard of the dashed black line and overlapping what might be considered the Antarctic continental margin. TAFZ indicates the Tasman-Australian fracture zone shown as a dotted, black line. It appears to disappear under the gravity high off the Oates Coast. The Balleny Islands (BI) are to the east of the Balleny fracture zone. TAS is the rotated and reconstructed position of TAS shown as a dashed white line, while the STR is shown reconstructed to the west of TAS as a dotted white line.

limit as the continental shelf break (CSB) and found a tight fit at 160 Ma between conjugate CSBs for parts of the southern margin of Australia with East Antarctica. Given the precedence established by *Royer and Sandwell* [1989], we adopt their nomenclature and use the term CSB. The BEDMAP sub-ice topography of East Antarctica [*Lythe et al.*, 2000] is shown in Figure 3 in the tight-fit Gondwanide reconstruction of *Lawver et al.* [1998] with East Antarctica held fixed in its present-day position. *Lawver et al.* [1998] used the CSBs determined from the satellite gravity data of *Sandwell and Smith* [1997] and the fit pole of rotation of *Royer and Sandwell* [1989] for Australia with East Antarctica. The reconstruction of Australia, Tasmania, and East Antarctica shown in Figure 3 is remarkably similar to the empirical fit of *Foster and Gleadow* [1992]. It is also very close to the fit of the reconstructed aeromagnetic data of *Finn et al.* [1999] who matched the magnetic signature between the Gleneig and Stawell zones of southeastern Australia with the magnetic signature of the Bowers zone of North Victoria Land, East Antarctica.

In Figure 3, good matches of the conjugate CSBs are found where continental East Antarctica is bordered by a sub-ice topographic high. This occurs along the margins of East Antarctica with respect to, NE Mozambique, Sri Lanka, the southern half of the eastern margin of India, and in particular, the region of Australia between 124°E and 133°E along the Great Australian Bight (GAB), shown in Figure 3 rotated against the Belgica Subglacial Highlands margin of East

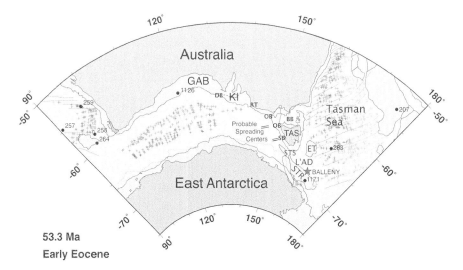

Figure 5. Polar stereographic plot of the reconstruction of AUS to EANT at the time of magnetic anomaly chron C24o (53.3 Ma). Location of a fixed BI hot spot is shown as a star. It may have produced the Eocene volcanics seen on the northeastern margin of the STR [*Exon et al.*, 2004; *Hill et al.*, 2001]. White is the region oceanward of the CSBs, taken as oceanic crust. Lighter gray crosses are magnetic anomaly chrons on the AUS plate, darker gray crosses are magnetic anomaly chrons on the Antarctic plate, and lightest gray crosses are picks from the Lord Howe Rise plate. Deep Sea Drilling Project (DSDP) and Ocean Drilling Program (ODP) sites are shown numbered. The present-day George V and Oates coasts margin is shown as a gray line west of the STR. The South Tasman Saddle immediately south of TAS may have been open to shallow to medium-depth water movement at early Eocene time. Abbreviations are ET, East Tasman Rise; L'A D, L'Atalante Depression. Basins shown along the southeastern Australian margin include BB, Bass Basin; DB, Duntroon Basin; OB, Otway Basin; RT, Robe Trough; SB, Sorrell Basin. There was no seaway south of the STR at this time although a few C24 anomaly picks are observed just south of the STR.

Antarctica between 120°E and 137°E. There are minor overlaps of the reconstructed conjugate CSBs, with one between India and East Antarctica (70°E to 85°E) and another between East Antarctica and the western section of the GAB (105°E to 120°E). These two minor overlaps are coincident with the Lambert Graben-Prdyz Bay Basin and the Aurora Subglacial Basin, respectively. Based on ODP Leg 119 drilling off Prdyz Bay [*Cooper et al.*, 1991], it is known that there are substantial glacially derived sediments prograded off the continental margin onto oceanic crust at Prdyz Bay. There may be as much as 200 km of Eocene? and younger sediments outboard of the original CSB in the Prydz Bay region. Along some Arctic margins are similar positive satellite gravity anomalies that *Vogt et al.* [1998] call Arctic marginal gravity highs. They related these gravity highs to recent Plio-Pleistocene North American glacial deposition on older oceanic crust that is not in isostatic equilibrium. The overlap found between the CSBs for closure of the South Atlantic in the region of the Niger delta has long been recognized [*Bullard et al.*, 1965], a case where deposition of riverine deposits are not in isostatic equilibrium with the oceanic crust and produce positive gravity highs. With the late Eocene to present timeframe for glacially derived deposition along the East Antarctic margin, lack of immediate isostatic equilibrium is a reasonable explanation for the overlap of the satellite gravity-derived CSBs where shown.

In contrast to the good match of the CSBs for the Antarctic margin from 120°E to 137°E with the western part of the (GAB) (124°E to 132°E for present-day Australian coordinates), there is a substantial and unacceptable overlap of the apparent Antarctic "continental margin" with the

Table 1. Euler Poles of Rotation for Australia to a Fixed East Antarctica

Anomaly	Age (Ma)	Latitude	Longitude	Angle	Reference
Present day	0.0	90.0	0.0	0.0	
An2ay	2.58	−16.2062	−141.3136	1.67	*Marks et al.* [1999]
An3y	4.24	−14.4472	−140.5404	2.73	*Marks et al.* [1999]
An3ay	5.89	−12.6122	−139.2836	3.84	*Marks et al.* [1999]
An4	7.75	−14.5478	−141.6163	5.02	*Marks et al.* [1999]
An5o	10.95	−11.4937	−141.5066	6.81	*Marks et al.* [1999]
An5ad	14.40	−11.2129	−142.7252	8.77	*Marks et al.* [1999]
An6y	19.05	−13.449	−145.339	11.48	*Marks et al.* [1999]
An13o	33.5	−13.45	−146.08	20.52	*Royer and Rollet* [1997]
An18o	40.1	−14.32	−148.25	23.21[a]	*Royer and Rollet* [1997]
An20o	43.8	−14.32	−148.25	24.32	this study
An33o	79.1	−3.87	−139.29	26.48	this study
An34y	83.0	−2.82	−138.41	27.15	this study
QZB	95.0	0.25	−135.87	27.88	this study
Fit	>125.0	2.0	−141.10	31.50	*Royer and Sandwell* [1989]

[a]Angle of rotation modified. Previously published poles used the *Cande and Kent* [1995] timescale.

Australian margin from 132°E to 141.5°E. The overlap on the Australian side consists in part of cratonic material, particularly in the region of Kangaroo Island. In Figure 4, the overlap of the rotated Australia outline is shown overlaid on the *Sandwell and Smith* [1997] satellite gravity data with Antarctica in its present-day position. If the "excess" Antarctic continental margin as defined by the satellite gravity gradient is removed from Antarctica off the Wilkes Land Basin region between 138°E and 158°E, then the amount of ocean crust formed between Australia and East Antarctica is compatible from the western edge of Australia at 115°E to the western margin of Tasmania at 144°E (Australia coordinates). Removal of the excess CSB as defined by the AMGH off the George V and Oates land coasts of East Antarctica leaves no need to surmise a radical change in plate geometry after breakup south of Australia. Simple closure works well west to east until the STR is reached (~150°E) where there appears to be less oceanic crust developed south of the STR than between central Australia and East Antarctica (Figure 5).

The STR continental fragments used in the paleoreconstruction figures are similar to the ones used by *Royer and Rollet* [1997] but reflect our own digitization of the steep gradient observed in the *Smith and Sandwell* [1997] satellite gravity data. The seismic refraction and reflection data of *Hinz et al.* [1990] suggest that much of the northern part of the STR has been stretched and extended by formation of Mesozoic and Cenozoic basins. We do not reduce the footprint of the STR to reflect the Mesozoic or possibly Cenozoic stretching, although a smaller, early Cenozoic footprint for the STR would affect the timing of the opening of any pre-Eocene seaway between the STR and East Antarctica. *Hinz et al.* [1990] suggested that the seismic reflection data show wrench faulting that affected the rift fill and transgressive strata up to the middle Eocene when they infer that early Cenozoic deformation stopped. They equated the end of wrench faulting with the final break between Australia and Antarctica, which supports a medium to deep passageway having formed between Australia and Antarctica as early as 40 Ma.

Major plate motions between Australia and East Antarctica are relatively well constrained for the Cenozoic [*Müller et al.*, 1997]. *Royer and Rollet* [1997] and *Cande and Stock* [2004a] found that seafloor spreading between Australia and East Antarctica started very slowly from the late Cretaceous (<5 mm yr^{-1}, half rate) until the middle Eocene, increased to about 15 mm yr^{-1} from magnetic reversal chron C20 (~43 Ma) to the beginning of C18 (~39 Ma) then increased to 24 mm yr^{-1} until ~33 Ma when the half-spreading rate increased again to 34 mm yr^{-1}. The change at C18 is consistent with other major plate reorganizations in the central Indian Ocean [*Royer*, 1992]. *Royer and Rollet* [1997] found slightly slower rates for the equivalent time periods between the STR and East Antarctica than at 142°E, immediately to the west of the STR. Since *Müller et al.* [1998] found an average of 3.1% asymmetric spreading between Antarctica and Australia with most of the asymmetry prior to 40 Ma, the difference in rates may simply be due to changing rates of asymmetry eastward along the ridge combined with some ridge jumps. The results of *Cande et al.* [2000], concerning possible Cenozoic extension in the Adare Trough, does not affect the time of opening of a Cenozoic seaway between Australia and Antarctica because the possible motion discussed in their paper is east of a seaway between the STR and East Antarctica.

The *Royer and Sandwell* [1989] pole for a tight-fit closure of Australia with a fixed East Antarctica is taken to be valid at ~160 Ma. In Table 1, the early poles of opening for Australia with respect to Antarctica until anomaly C13 (33.1 Ma) have been modified from the work of *Royer and Rollet* [1997] using the magnetic anomaly picks of *Tikku and Cande* [2000]. The *Marks et al.* [1999] poles of rotation for C6 (~19.0 Ma) to present are used as published and are similar to the poles in the work of *Cande and Stock* [2004b]. Other than the question of what constitutes the continental crust of East Antarctica between 138°E and 158°E, there is little controversy about the precise positions of East Antarctica with respect to Australia for the Cenozoic.

3. RECONSTRUCTION OF AUSTRALIA WITH EAST ANTARCTICA

The most significant overlap for a tight-fit between the CSBs of East Antarctica and Australia is shown hatched in Figure 3 between 138°E and 158°E for Antarctic longitudes. This overlap includes the area of Kangaroo Island off Australia (Figure 4) where the geology is well known [*Foden et al.*, 2002]. The excellent match of the CSBs to the west implies that the overlap is real. We assume that the problem lies on the relatively unknown Antarctic plate and not on the Australian plate since there is no known postbreakup extension on the continental portion of either plate between the area of good agreement (120° to 135°E) and the area of substantial overlap. The conjugate Antarctic region to Kangaroo Island is offshore of the Wilkes Land Basin (Figure 3) and lies between the Belgica Subglacial Highlands (B on Figure 3) and the Transantarctic Mountains (T on Figure 3). The apparent protuberance along the Antarctic coast at 145°E is the Mertz Glacier Tongue and hence not indicative of continental material. The deposition of glacial material that occurred outboard of the shelf break, producing the extended CSB, came from the Cook, Ninnis, and Mertz glaciers (see *Rignot et al.* [2008] for locations of the glacial drainage basins). The western edge of this overlap zone was drilled by IODP Leg 318 [*Expedition 318 Scientists*, 2010], where they found as much as 602 m of middle Miocene and younger sediments at a seafloor depth of 4002 m.

If the overlap zone of the margins of George V Land and Oates Land between 138°E to 158°E is post-Eocene in age, then determination of the seaward extant of the true continental crust has tremendous implications for the timing of the opening of a seaway south of the STR. In Figure 4, the East Antarctic CSB anomaly appears to overlie the southern end of the Tasman-Antarctic fracture zone (TAFZ), which seems to disappear beneath the CSB, leaving an irregular margin that between 66°S and 68°S should be roughly subparallel to the Balleny Fracture Zone to the east (Figure 4). The rotated, present-day coastline of Australia (shown as thicker white line in Figure 4) overprints the CSB of the Oates Land margin between 154°E and 158°E. On the left side of Figure 4, west of 138°E, the rotated Australian CSB overlies precisely the CSB of East Antarctica (shown as solid black line). This excellent match extends westward to 120°E as seen on Figure 3. A rotated Kangaroo Island (KI on Figure 4) overlies the black-dashed CSB of East Antarctica and nearly overlies the Merz Glacial tongue. Therefore, we assume that the rotated Australian CSB might be taken as a proxy for the actual East Antarctic CSB between 138°E and 158°E. Consequently, the space between the rotated Australian CSB and the dashed black line is assumed to be the outline of the glacially derived material transported by the East Antarctic Ice Sheet.

The apparent narrowing of oceanic crust formed eastward along the Australian-Antarctic ridge from 130°E to 160°E has led some to surmise a complicated eastward propagating rift between Australia and Antarctica, first suggested by *Mutter et al.* [1985]. Removal of the excess, apparent continental shelf off the George V Land and Oates Land coasts eliminates the need for the presumption of an eastward propagating spreading center since the space shown in the reconstruction at 53.3 Ma (Figure 5) indicates little need for differential spreading south of western Australia versus immediately west of Tasmania. The 53.3 Ma reconstruction does suggest that there was a northward ridge jump shortly before the Eocene in the area immediately to the south of Kangaroo Island.

Wilcox and Stagg [1990] suggested initiation of extension in the GAB as early as 153 Ma. How spreading occurred between the tight-fit shown in Figure 3 and the early Eocene reconstruction shown in Figure 5 is not germane to the opening of a seaway between the STR and East Antarctica but will be noted here briefly. The initial extension in the GAB extended to the east [*Wilcox and Stagg*, 1990] and ran first into the Duntroon Basin and Robe Trough with transtension in the Otway Basin northwest of Tasmania (for locations, see Figure 5). At ~120 Ma, the initial NW-SE extension changed to NE-SW extension with opening in the Otway, Sorell, and Bass basins until ~95 Ma. *Wilcox and Stagg* [1990] suggest that true seafloor spreading commenced between Australia and East Antarctica sometime after 95 Ma. The Quiet Zone Boundary (QZB) represents the first true oceanic crust [*Wilcox and Stagg*, 1990]. *Tikku and Cande* [2000] also picked a magnetic trough or "MT" anomaly and suggest that some seafloor spreading may have occurred between the QZB and the MT. *Tikku and Cande* [2000] make the point that the QZB is not necessarily the Ocean-Continent boundary since there is clearly a mixture of oceanic and continental crust in this zone based on the seismic refraction results of *Talwani et al.* [1979]. *Talwani et al.* [1979] used the term Magnetic Quiet Zone or MQZ on which the QZB is based because they thought that C33/34 was actually C22 and that the anomalies older than C22 were missing and therefore abnormally "quiet," hence MQZ. *Cande and Mutter* [1982] revised the magnetic anomaly picks south of Australia and reidentified C22 as C33/34. *Veevers* [1986] then calculated the age of 95 Ma for the QZB based on the very slow spreading rate determined by *Cande and Mutter* [1982] for anomalies C20 through C34. In fact, as illustrated by *Tikku and Cande* [2000], the MQZ is not particularly quiet except on their easternmost profile at 130°E. *Stagg et al.* [1999] leave open the possibility that the period of slow seafloor spreading between Australia and Antarctica commenced much earlier than the Cenomanian as suggested by *Cande and Mutter* [1982] and suggest that seafloor spreading may have commenced in the Neocomian or Late Jurassic and was active along much of the southern margin of Australia at the earlier time.

4. SOUTH TASMAN SADDLE

For purposes of timing the opening of a seaway south of Australia, it is important to determine the age of seafloor spreading or extension in the South Tasman Saddle. If the magnetic anomalies, tentatively identified by *Royer and Rollet* [1997] as chrons C33 to C30 (79 to 66 Ma) [*Gee and Kent*, 2007], to the east of the South Tasman Saddle in the L'Atalante Depression (Figure 5) are correct, and can be taken as the time of stretching and seafloor spreading between Tasmania and the STR, then the South Tasman Saddle developed during the Campanian to Maastrictian. By early Eocene, seafloor spreading in the Tasman Sea had ceased [*Gaina et al.*, 1998] and the magnetic anomalies, C24 and younger created at the Australia-Antarctic ridge can be clearly identified; see the picks from *Tikku and Cande* [2000] shown as the C24 (53.3 Ma) picks in Figure 5. *Boreham et al.* [2002] note that there was an accelerated rate of oil generation at ~48 Ma in the Sorell Basin, in response to the maximum burial heating rate in the early Eocene which coincides with the passage of the nearby spreading center as shown in Figure 5. Chron 33/34 is perhaps the only clearly identifiable anomaly on the Australian side older than C24 and is seen on four of the five lines that *Tikku and Cande* [2000]) highlight in their Figure 4. While we use the picks from *Tikku and Cande* [2000], we have calculated our own poles of rotation (Table 1) based on interactively matching the picks for each identified magnetic anomaly. We interpolate between those times and assume steady state spreading between anomaly picks in order to create the reconstructions shown.

From 40 to 30 Ma, the active Australia-Antarctic spreading center migrated past the western end of the South Tasman Saddle. The increased thermal input may have elevated the Saddle such that a significantly deep seaway between the STR and Tasmania was precluded. Prior to any influence of the spreading center, the South Tasman Saddle may have been a shallow to medium-depth seaway between Australia and East Antarctica in the late Paleocene to early Eocene. If the George V/Oates Coast shelf break was substantially closer to the present East Antarctic shoreline, then a deep seaway to the south of the STR may have existed as early as 40 Ma. Only if the wider George V/Oates Coast margin was in place prior to the late Eocene and the South Tasman Saddle was not sufficiently deep to allow substantial circulation, would a medium-depth to deepwater, Australian-Antarctic seaway, have developed only after late Eocene as suggested by *Lawver and Gahagan* [2003]. Whenever a passage opened between Australia and East Antarctica, vigorous, clockwise-circumpolar transport of water masses may not have begun until a deepwater opening south of South America completed the high latitude, circum-Antarctic seaway or even later depending on global ocean circulation paths. The initial, gradual increase in ∂O^{18} immediately after the EECO, and the presumed cooling of ocean temperatures, was most probably caused by the closure of the Tethyan seaway north of India. The initial collision of greater India with Eurasia diverted the circumtropical worldwide circulation that went north of Africa at ~30°N, to a circumtemperate circulation to the south of Africa at ~45°S [*Lawver and Gahagan*, 1998]. Major transport of water by a true ACC may not have developed until the Miocene closure of equatorial and tropical seaways increased anticyclonic gyres in the temperate zones which, in turn, increased forcing of the ACC [*Lawver and Gahagan*, 1998].

From the reconstruction shown in Figure 5, it is apparent that the South Tasman Saddle cleared East Antarctica well before the early Eocene, whether or not the CSB off the George V/Oates Coast of East Antarctica is the dotted line or the solid line. If the stretching and seafloor spreading in the South Tasman Saddle formed between C33y (73.6 Ma) and C30y (65.6 Ma) as suggested by *Royer and Rollet* [1997], then the seafloor there would have been approximately 20 Myr old by the beginning of the Eocene and perhaps at least 1000 m deep. The asymmetric position of C24 (53.3 Ma) between 130°E and 140°E (Figure 5) to the west of Tasmania and south of Kangaroo Island indicates that a northward ridge jump occurred in this area just prior to the Eocene or at least by C24 time. The northward jump left most of the older, pre-C24 seafloor on the Antarctic plate for the region immediately to the west of Tasmania. There is a slight suggestion in the satellite-derived gravity data off the East Antarctic margin (Figure 4) of what may be an abandoned spreading center at 63°S between 133°E and 141°E off the Adélie Coast. If the AMGH region is eliminated from both the George V and Oates coasts, then the hypothesized, late Cretaceous magnetic anomalies under the George V Coast AMGH would nearly align with the proposed C33? to C30? magnetic anomalies in the L'Atalante Depression immediately to the east of the South Tasman Saddle.

Willcox and Stagg [1990, Figure 9] show an initial 300 km of NW-SE extension between Australia and East Antarctica transformed to the south of the STR along the western margin of the STR. This initial extension was followed by later seafloor spreading between Australia and East Antarctica to the west of Tasmania. To the east, Cretaceous stretching began first between Tasmania and Australia initiating the Otway and Bass basins and later by the stretching between Tasmania and the STR producing the South Tasman Saddle. Finally, the plate boundary switched to seafloor spreading south of the STR.

5. HOT SPOT ACTIVITY

An active, late Cretaceous Balleny Islands hot spot [*Duncan and MacDougall*, 1989; *Lanyon et al.*, 1993] may have had a major influence on the region of the East and South Tasman rises. In an absolute plate motion framework based on "fixed" Indo-Atlantic hot spots [*Müller et al.*, 1993], plate reconstructions indicate that a fixed Balleny Islands hot spot may have initiated opening of the Tasman Sea as postulated by *Lanyon et al.* [1993]. The start of the opening of the Tasman Sea is dated at 87 Ma by *Gaina et al.* [1998] and may mark the beginning of the hot spot activity. The East Tasman Rise would have been above a fixed Balleny Islands hot spot only from 85 to 80 Ma. The hot spot would have been between the East Tasman Rise and Tasmania from 82 to 70 Ma and beneath the L'Atalante Depression from 70 to ~53 Ma. It may have produced the numerous small volcanoes shown on the southeast margin of Tasmania [*Crawford et al.*, 1997, Figure 2] and is a good explanation for the prominent seamounts in the L'Atalante Depression shown in Figure 6 of the work of *Exon et al.* [1997a, 1997b] and in Figure 1 of the work of *Hill et al.* [2001]. The suggested motion of the East Tasman Rise, with

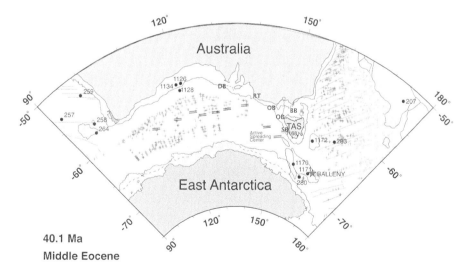

Figure 6. Polar stereographic plot of the reconstruction of AUS to EANT at the time of chron C18o (40.1 Ma). Location of a fixed BI hot spot is shown as a star. White region is oceanward of the CSBs. Medium gray crosses are magnetic anomaly chrons on the AUS plate, darker gray crosses are magnetic anomaly chrons on the Antarctic plate, and lightest gray crosses are picks from the Lord Howe Rise plate. DSDP and ODP sites are shown numbered. If the Oates Coast shelf was the dashed line shown to the west of the STR, then the first seaway to the south of the STR may have begun open to deep water transport at this time.

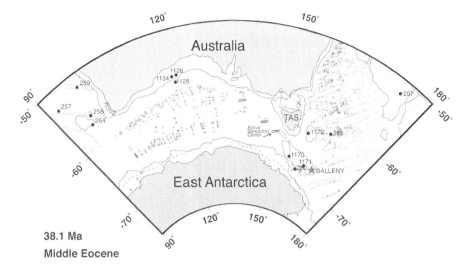

Figure 7. Polar stereographic plot of the reconstruction of AUS to EANT at the time of chron C17o (38.1 Ma). Symbols and abbreviations as in Figures 5 and 6. The active AUS-Antarctica spreading center would have just begun to affect the western end of the South Tasman Saddle. The Balleny hot spot may have been active during this period and may have produced the southern third of the STR as a strictly Cenozoic construction.

respect to Tasmania and the opening of the South Tasman Saddle between the STR and Tasmania, generally follow the scenario suggested by *Exon et al.* [1997a, Figure 6]. If the hot spot initiated the opening between the East and South Tasman rises, then its track would also support the magnetic anomaly identifications of C33y? to C30y? that *Royer and Rollet* [1997] identified in the L'Atalante Depression.

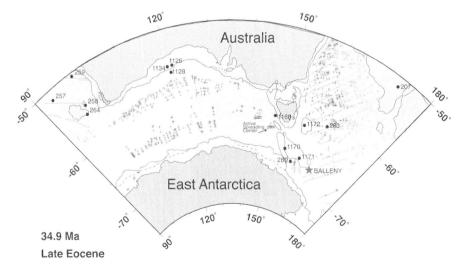

Figure 8. Polar stereographic plot of the reconstruction of AUS to EANT at the time of chron C15o (34.9 Ma). Symbols are as in Figures 5 and 6. The active AUS-Antarctica spreading center may have been directly off the western end of the South Tasman Saddle. The STR was no longer above the Balleny hot spot, and the hot spot may have begun a period of diminished or no activity. The broader Oates Coast continental shelf, if it existed, would have blocked a seaway to the south of the STR.

In the reconstruction for 53.3 Ma (Figure 5) and on subsequent reconstruction figures of the Australia-Antarctica region, the location of a fixed Balleny Island hot spot is shown as a star. In an absolute reference frame, the "Volcanic (?hot-spot)" location just to the northeast of the STR described by *Hill et al.* [2001] and shown on Line C in their Figure 7 appears nearly stationary over the hot spot from 62 to 51 Ma. The calculated hot spot track indicates that the eastern margin of the STR would have been above a fixed Balleny hot spot from ~51 until ~35 Ma and would account for the Eocene volcanic rocks dredged along the eastern scarp of the STR, shown by *Hill et al.* [2001] on their seismic profiles C and D across the STR. There is no convenient "hot spot trail" that connects the STR to the Balleny Islands particularly once the Australian plate began to move rapidly northward in an absolute reference frame. There is a feature in the satellite gravity data of *Sandwell and Smith* [1997] at 52°20'S, 152°40'E, approximately 300 km SSE of the STR that may be on a projected hot spot track between the STR and the present-day Balleny Islands. As noted by *Lanyon et al.* [1993], it is only after the Australia-Antarctica spreading center passed over the hot spot during the Miocene that the Balleny Islands themselves begin to form. The Balleny Islands are on seafloor dated as young as 10 Ma, but the southern part of the Balleny Islands ridge may have been over the hot spot beginning about 20 Ma. A fixed Balleny Islands hot spot is a worthy candidate for a mechanism to uplift the whole STR region until just before the Eocene-Oligocene boundary.

6. MIDDLE EOCENE

By middle Eocene (40.1 Ma, $C18_o$), there was a gap between the STR and East Antarctica (Figure 6) if the region of the AMGH off the George V and Oates coasts is a postbreakup Oligocene glacial feature. In addition to any possible seaway south of the STR, there is the possibility that a medium to shallow seaway through the South Tasman Saddle existed at this time. The active Australian-Antarctic Ridge spreading center was still to the north of the western end of the South Tasman Saddle. If it was active, a fixed Balleny Islands hot spot would have been

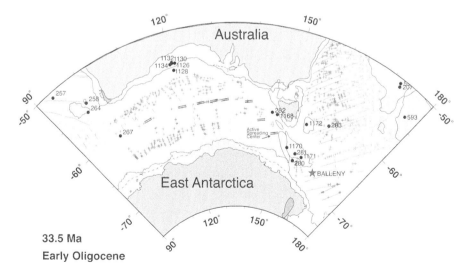

Figure 9. Polar stereographic plot of the reconstruction of AUS to EANT at the time of chron C13o (33.5 Ma). Symbols are as in Figures 5 and 6. The active AUS-Antarctica spreading center would have been south of the western end of the South Tasman Saddle at this time. The broader Oates Coast margin would just barely block a seaway to the south of the STR.

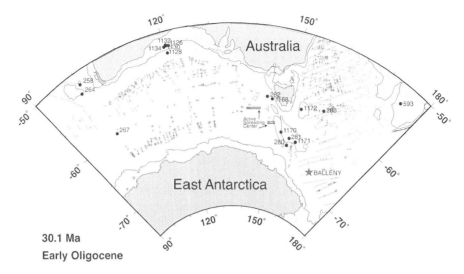

Figure 10. Polar stereographic plot of the reconstruction of AUS to EANT at the time of chron C11o (30.1 Ma). Symbols are as in Figures 5 and 6. By 30 Ma, there had to have been a deepwater passage to the south of the STR, no matter where the Oates Coast continental shelf is chosen. At the time shown, there is a positive anomaly on the satellite altimetry data that suggests a seamount at the location of a fixed BI hot spot.

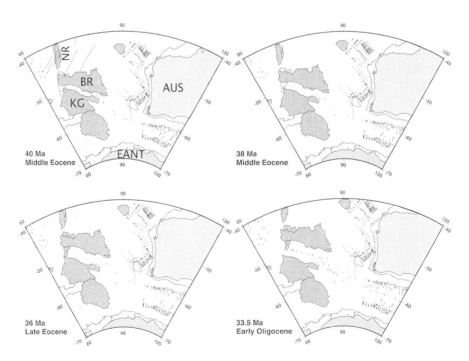

Figure 11. Polar stereographic plots of reconstructions of the opening between KP and BR for the times 40, 38, 36, and 33.5 Ma. Symbols are as in Figures 5 and 6. Abbreviations are KG, Kerguelen; NR, Ninety East Ridge.

under the southeastern margin of the STR, immediately to the east of ODP Site 1171 as shown in Figure 6. In Figure 7 (38.1 Ma), the magnetic anomalies immediately to the west of Tasmania as identified by *Royer and Rollet* [1997] determine the location of the active Eocene spreading center, and the STR had moved northward with respect to a Balleny Islands hot spot. The lack of definitive structural data constraining the nature of the southern end of the STR leaves open the possibility that the STR south of 49°S may simply be a product of an active hot spot and not continental. Therefore, as a bathymetric high, the southern end of the STR may not be older than middle Eocene.

7. LATE EOCENE

By 34.9 Ma (Figure 8), the Australian-Antarctic Ridge spreading center would have been off the western end of the South Tasman Saddle and may have uplifted the seaway slightly to lessen or eliminate any major throughput, although the lateral heat flow was probably not significant so the potential uplift may have been slight. By the start of the late Eocene (37.0 Ma), the southern end of the STR had probably passed over a fixed Balleny Islands hot spot, and if the STR had been thermally uplifted by the hot spot, it would have begun to subside, possibly rapidly. If the AMGH represented by the shelf area between the solid and dotted lines off the Oates coast of East Antarctica represents material that was in place prior to the late Eocene, then a seaway south of the STR would have been closed to deep water flow. If the AMGH was emplaced after the end of the Eocene, then there would have been a 200+ km wide, deep water seaway to the south of the STR.

8. EARLY OLIGOCENE

By anomaly chron C13 time (33.5 Ma) shown in Figure 9, the active Australian-Antarctic Ridge would have passed the western end of the South Tasman Saddle. As noted above, if the Oates Coast shelf is an Oligocene or younger feature, there would have been a substantial seaway to the south of the STR. The general lack of seafloor features immediately to the south of the STR suggests that the hot spot may have gone into a quiescent period similar to what happened to the Hawaii-Emperor hot spot between about 41 and 25 Ma. By 30.1 Ma (Figure 10), a significant seaway is open to the south of the STR whether or not there is a wide continental margin along the George V/Oates coasts of East Antarctica. As mentioned above, the gravity feature located at 52° 20′S, 152°40′E would have been above a Balleny Islands hot spot at about 30 Ma although it is of unknown origin. *Stickley et al.* [2004] defined four phases of opening of the Tasmanian Gateway with their Phase C lasting from 33.5–30.2 Ma, which included deepening of the gateway to bathyal depths, with episodic erosion by increasingly energetic bottom-water currents. Their Phase D, defined as younger than 30.2 Ma, resulted in the establishment of stable, open-ocean, warm-temperate, oligotrophic settings in the vicinity of the STR characterized by siliceous-carbonate ooze deposition. *Stickley et al.* [2004] conclude that the initial deep, early Oligocene Tasmanian Gateway produced an eastward flow of relatively warm surface waters from the Australo-Antarctic Gulf into the southwestern Pacific Ocean. Their "proto-Leeuwin" current was decidedly not the same as an eastward flowing "proto-ACC" during the early Oligocene and therefore the ACC as presently known was a later phenomena.

9. KERGUELEN PLATEAU–BROKEN RIDGE

A potential blockage to major west-to-east circulation south of Australia may have been the final opening between Broken Ridge and the northern margin of Kerguelen Plateau. While

Broken Ridge and Kerguelen Plateau separated at about 43 Ma, the southern end of 90° East Ridge may have acted as a barrier to open circulation until sometime after 36 Ma (Figure 11). Even if one of the two earlier options for an open Tasmanian Gateway did produce a significant seaway, there may have been little or no transport through such a seaway if there was no global force to drive circulation or if Drake Passage was not open to allow complete circum-Antarctic circulation.

10. OPENING OF DRAKE PASSAGE

The evidence for the time of the opening of Drake Passage is not simple. There are a number of small continental fragments in the Scotia Sea whose paleopositions are not constrained because the magnetic anomalies needed to properly place them in their correct paleopositions are either lacking or not easily correlated and interpreted. By Eocene, all motion between East and West Antarctica in the Weddell Sea region had ended. Consequently, the major plate framework for the Drake Passage region can be deduced from the motion between South America (SAM) and Africa (AFR) and between Africa (AFR) and an East Antarctica (EANT) assumed fixed to West Antarctica (WANT). For our work, we use rotation poles for SAM-AFR from the work of *Müller et al.* [1999], for AFR-EANT for 85 to 65 Ma from the work of *Royer et al.* [1988], for 65 to 20 Ma from the work of *Royer and Sandwelli* [1989] and for 20 Ma to present from *Royer and Chang* [1991]. In addition to the major plate motion, we assume that Powell Basin opened between 34 and 30 Ma based on heat flow versus age comparisons as discussed in the work of *Lawver and Gahagan* [1998]. *Coren et al.* [1997] dated Powell Basin as 27 to 18 Ma based on a magnetic anomaly correlation of one line with highly asymmetrical spreading and amplitudes of the magnetic anomalies of about 100 nT, while *Eagles and Livermore* [2002] dated seafloor spreading in Powell Basin as 29.8 to 21.8 Ma with more symmetrical spreading but no real correlation of anomalies between lines. For comparison, *Livermore et al.* [2000] show an excellent correlation of magnetic anomalies along the extinct Phoenix-Antarctic Ridge immediately to the west of Drake Passage with magnetic anomaly amplitudes approaching 500 nT with clear definition of anomalies chrons 5AC (14.1 Ma) to C2An.3n (~3.6 to 3.3 Ma). Heat flow versus age correlations in Jane Basin [*Lawver et al.*, 1991] give an opening age of 32 to 25 Ma. The youngest magnetic anomalies found immediately to the east of Jane Bank [*Barker et al.*, 1984] are unequivocally chron C6a (~21 Ma) or older and imply that Jane Basin may have opened as a back arc basin. While this scenario is in agreement with *Maldonado et al.* [1998], our model suggests ages for Powell Basin and Jane Basin about 10 million years older than they use.

Comparison of the geology of South Georgia and the southern tip of South America requires pre-breakup placement of South Georgia to the east of Tierra del Fuego [*Dalziel*, 1981], although an initial location of South Georgia at the eastern end of Burdwood Bank would produce a less complicated model. A location of South Georgia adjacent to Tierra del Fuego produces a limited number of plausible modes of opening of Drake's Passage during the middle to late Tertiary. Both *Barker and Burrell* [1977] and *Eagles et al.* [2005] identified seafloor spreading anomalies in the western Scotia Sea as old as chron C8 (~26.5 Ma), while *LaBrecque and Rabinowitz* [1977] show a chron C10 (28.5 Ma). *Lodolo et al.* [1997] suggest an area in the southwest corner of the western Scotia Sea with an age older than anomaly C10 based on geophysical data. *Eagles et al.* [2005] updated the anomaly picks in the western Scotia Sea and suggested inception of seafloor spreading prior to chron C8 (26.5 Ma) and extinction around chron C3An.2n (6.6–5.9 Ma). While they show a few chron ?C9 (28.0–27.0 Ma), they do not recognize the chron C10 of *LaBrecque and Rabinowitz* [1977].

Barker [1995] shows a Drake Passage clogged with continental blocks well into the Miocene (20 Ma) [*Barker*, 1995, Figure 7.2c]. In his review paper, *Barker* [2001] shows a nascent Scotia Sea extant at chron C13y (33.01 Ma), but he does not show a fully open Drake Passage until chron $C6_o$ (20.13 Ma). From his figures, it appears that the Shackleton Fracture Zone Ridge (SFZR) blocked an open Drake Passage until sometime between chrons A8o and A6o perhaps finally clearing the southern tip of the South American continental shelf about 23 Ma. At 20 Ma, he has a very tight closure around the eastern margin of the Scotia Sea and suggests that a more compact north Scotia Ridge may have prevented continuous circumpolar deepwater flow. *Barker* [2001] assumed that the (SFZR) that is as shallow as 700 m for much of its length was a feature that predated opening of Drake Passage. The ridge is shown on Figure 3E of the work of *Barker* [2001] for chron 13y (33.01 Ma) and appears to block any deepwater throughput at an initial Drake Passage at $C8_o$ (26.55 Ma). Both *Klepeis and Lawver* [1996] and *Livermore et al.* [2004] argued that the (SFZR) is, in fact, a recent construct, formed by underthrusting of the Scotia plate under the Antarctic plate, a theory supported by *Galindo-Zaldivar et al.* [2000]. *Barker* [2001, Figure 3B] does show a nascent Shag Rocks passage for chron C5C (16 Ma).

Hill and Barker [1980] show obviously lineated east-west trending magnetic anomalies in the central Scotia Sea. Their correlation of the age of the marine magnetic anomalies is not convincing, but it can be presumed that they predate the identified anomalies in the east Scotia Sea that *Barker* [1995] identified as chron C5a (~12 Ma) and possibly even as old as chron C5b (14.8 Ma). *Larter et al.* [2003] updated the earlier work and confirmed seafloor spreading as early as C5b and possibly before chron C5c (16.1 Ma). *Hill and Barker* [1980] tentatively identified the central Scotia Sea anomalies as chron C6 (19.6 Ma) to as young as "6 Ma" based on extrapolation of a 10 mm yr^{-1} spreading rate after 16 Ma. *Maldonado et al.* [2003] recently identified magnetic anomalies chrons C6a (21.2 Ma) to C5AD (14.4 Ma) along one line slightly to the east of the *Hill*

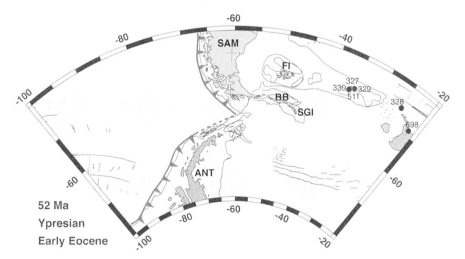

Figure 12. Polar stereographic plot of the reconstruction of the DP between SAM and the Antarctic Peninsula (ANT) at 52 Ma, early Eocene. This time represents the most probable last time that mammals were able to migrate freely between SAM and Seymour Island on the ANT. South Georgia is shown in an unconstrained position just south of BB, while the plate locations are based on major plate motions as discussed in the text. Magnetic anomalies are shown in black, either as dots for individual picks or lines as isochrons. DSDP and ODP holes are labeled. The subduction zone is shown with the teeth pointing in the direction of the downgoing slab. The gray area on the eastern margin of the figure represents the Large Igneous Province of the Northeast Georgia Rise.

and Barker [1980] anomalies. Although *Maldonado et al.* [2003] do not provide their magnetic correlation model with the spreading rates they used, the magnetic anomalies they show would be compatible with the ages identified by *Barker* [1995] for the oldest east Scotia Sea anomalies. *Maldonado et al.* [2003] do indicate that their anomaly identifications are offset from the *Hill and Barker* [1980] anomalies, but the presumed seafloor spreading in the central Scotia Sea still terminates prior to that in the east Scotia Sea.

The age of the central Scotia Sea is questioned by *Eagles et al.* [2005] who suggest that it must be older than both the western and eastern Scotia seas. They analyzed the magnetic anomalies for the west Scotia Sea and found that the simplest explanation for the apparent coupling of seafloor spreading at both the west Scotia Ridge and east Scotia Ridge with trench migration and South America-Antarctica motion is that the central Scotia plate acted as an older "arc" plate between C8 and about C5C, when the east Scotia Sea spreading center was initiated. Since there is no evidence for an abandoned spreading center in the central Scotia Sea as the scenario suggested by *Hill and Barker* [1980] implies, it is assumed that the anomalies in the central Scotia Sea are older than chron C8 as suggested by *Eagles et al.* [2005]. Whether the anomalies assigned to the Dove and Protector basins by *Eagles et al.* [2006] are, in fact, 41 to 34.7 and 34 to 30 Ma, respectively, as they conclude, cannot be determined by the very short magnetic reversal sequences found in the two basins. More recently, *Eagles* [2010a] has dated the magnetic anomalies in the central Scotia Sea as being Jurassic to Early Cretaceous in age, from chron M25 (154 Ma) to possibly chron M1 (121 Ma), although the correlation shown is not persuasive, and he does not have a good explanation for the depth of the central Scotia Sea, which should be considerably deeper if that old. What can be concluded is that given the constraints on the suggestion of the predrift placement of South Georgia just off Tierra del Fuego [*Dalziel et al.*, 1975], the first identified

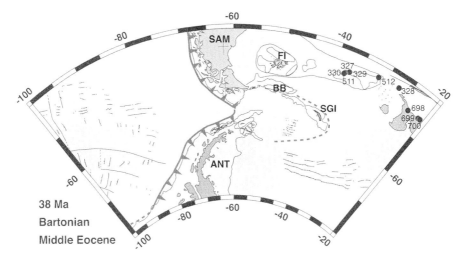

Figure 13. Polar stereographic plot of the reconstruction of the DP between SAM and the ANT at 38 Ma (middle Eocene). In comparison with Figure 12, it is clear that the tip of the ANT has moved to the east with respect to SAM. While there are few real constraints on the motion of South Georgia, it is shown now to the east of BB. Incipient westward directed subduction is indicated by the dashed line that encircles South Georgia as it moves to the east with respect to SAM. Abbreviations are as in Figure 12. Magnetic picks to the south and west of South Georgia are the ones from the central Scotia Sea, presumed to be older than the ones identified in the west Scotia Sea by *Eagles et al.* [2005]. The region between the ANT, BB, and to the west of South Georgia is assumed to have been produced by the westward motion of SAM with respect to the Antarctic plate with the central Scotia Sea/South Georgia block moving with the ANT.

magnetic anomalies in the western Scotia Sea, and the direction of Oligocene to Miocene seafloor spreading in the western Scotia Sea, there are few viable scenarios to translate South Georgia from just east of Tierra del Fuego to its present position over 1000 km to the east.

Eagles [2010b] recently suggested an initial location of South Georgia to the east of Burdwood Bank, but such a location is not supported by the mapped geology of South Georgia. He felt that it is impossible to find evidence in the ocean floor geophysical data for the full translation of South Georgia to its present position and therefore makes the assumption that such a translation could not have occurred. He goes on to develop an alternative scenario that the South Georgia microcontinent originated "within the interior of Gondwana," adjacent to the southern margin of the Maurice Ewing Bank at the eastern extremity of the Falkland Plateau. His scenario requires that all the igneous and deformational events recorded in the rocks of South Georgia took place in isolation several degrees of latitude north of the Andean Cordillera in Tierra del Fuego with which its geology matches so well. In contrast, the physiography and geology of southeastern Tierra del Fuego clearly indicate that a continental fragment exactly the size and shape of the South Georgia microcontinent has been removed from the margin east of Isla Navarino and south of Isla de los Estados and of Burdwood Bank. The missing continental fragment is not found in the Antarctic Peninsula or elsewhere to the south. The highly deformed rhyolitic igneous rocks of the Upper Jurassic Tobifera Formation of southern South America strike eastward along the length of Isla de los Estados and out to sea along the southern margin of Burdwood Bank where seismic refraction data [*Ludwig et al.*, 1968] indicate they continue to its eastern extremity. Given that there is a missing fragment of the Andean Cordillera that exactly matches the scale, geological structure, and history of the South Georgia microcontinent, an extra-Andean setting seems far less likely than the Pacific margin hypothesis put forth by *Dalziel et al.* [1975].

For the middle Eocene (Figure 12, 52.0 Ma), major plate motions between SAM-AFR and AFR-ANT result in the tip of the Antarctic Peninsula placed against the southern margin of Tierra

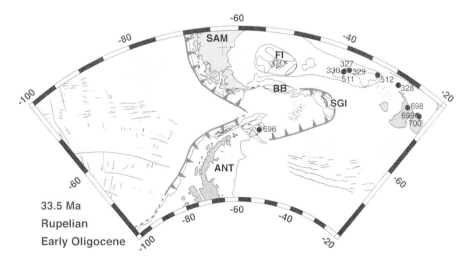

Figure 14. Polar stereographic plot of the reconstruction of the DP between SAM and the ANT at 33.5 Ma (Chron 13o, earliest Oligocene). Without good constraints on the motion of South Georgia, it is difficult to determine the precise extent or motion of the older Scotia Sea. There is some motion beginning to happen in the Powell Basin, but the precise age of opening is not known. Westward dipping subduction to the east of South Georgia must be occurring by this time as South Georgia moves with respect to the Falkland Plateau and the South American plate.

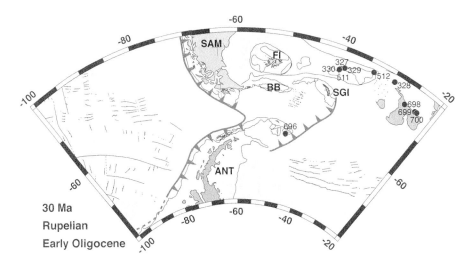

Figure 15. Polar stereographic plot of the reconstruction of the DP between SAM and the ANT at 30 Ma (Chron 11o, early Oligocene). There are still no identified anomalies in the west Scotia Sea. Based on heat flow measurements in Powell Basin [*Lawver and Gahagan*, 1998], the Powell Basin may have been fully open by this time, and Jane Basin to the east of the SO was beginning to open, see Figure 2 for locations of Powell and Jane basins.

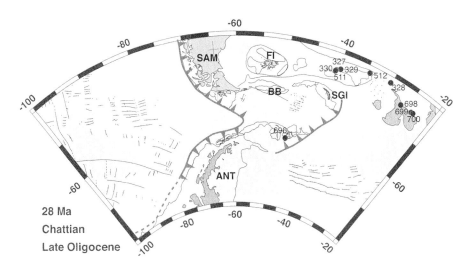

Figure 16. Polar stereographic plot of the reconstruction of the DP between SAM and the ANT at 28 Ma (late Oligocene). The oldest identified magnetic anomalies in the Scotia Sea are shown off Tierra del Fuego, taken from the work of *LaBrecque and Cande* [1985]. By late Oligocene, there are insufficient continental blocks to prevent deepwater flow through Drake Passage. The Shackleton fracture zone ridge did not exist until Pliocene [*Klepeis and Lawver*, 1996], so the ACC could have been established by this time. If there was a medium-depth arc that extended along the subduction zone between South Georgia Island and the SO near ODP Hole 696, then a vigorous ACC may not have developed until the Shag Rocks Passage opened. See Figure 2 for location of the Shag Rocks Passage, although it may not have subsided below 2500 m until middle to late Miocene time.

del Fuego, allowing no room for a deep water Drake Passage. In addition, it is known that mammals were able to migrate between Antarctica and South America at this time. Neither the exact location of the South Georgia microcontinent nor the precise tectonic regime is known for this time, but it can be assumed that the Phoenix plate was being subducted beneath South America and Antarctica in the early Eocene [*Eagles et al.*, 2004]. By late middle Eocene (Figure 13, 38 Ma), major plate motions predict that the Antarctic Peninsula had begun to move eastward with respect to the tip of South America [*Cunningham et al.*, 1995] perhaps having generated some of the crust that presently lies in the southern part of the west Scotia Sea. Depending on the age of the Shackleton Fracture Zone Ridge and the timing of the opening of the Powell Basin, Drake Passage may have still been closed to deepwater flow but it is quite probable that shallow flow was possible by the end of the middle Eocene. *Scher and Martin* [2004] present ε_{Nd} values that suggest transport of Pacific water masses into the Agulhas Ridge region by 41 Ma. While the motion of South Georgia from where it was off Tierra del Fuego in the Cretaceous to where it must have been in the Oligocene is unknown, we show some eastward motion of the South Georgia block during the Eocene as that is what is required such that the identified chron C8 found off the eastern end of Burdwood Bank can be reconstructed with the ones found abutting the central Scotia Sea block. While it may have initially moved eastward as part of the Antarctic plate as implied by *Eagles et al.* [2005], in order for there to be sufficient room for some of the continental fragments identified by *Barker and Burrell* [1981], it is moved independently until 38 Ma (Figure 13).

As the South American plate moved westward with respect to the Antarctic plate and with respect to South Georgia, a nascent west Scotia Sea must have opened. This implies that westward directed subduction was occurring beneath the central Scotia Sea region. By 33.5 Ma, a nascent western Scotia Sea had opened as a result of major plate motions, but Drake Passage per se may have still been blocked to deepwater flow by elevated fragments including Burdwood Bank, South Georgia, possibly with the central Scotia Sea block, and Discovery Bank (Figure 14). *Barker* [2001] shows a reconstruction for the Scotia Sea region for the Oligocene that is similar to the one shown in Figure 14, with the exception that he has a (SFZR) in place at 33 Ma. We assume though that the SFZR is a recent feature that resulted from transpression between the Antarctic and Scotia plates [*Klepeis and Lawver*, 1996] once seafloor spreading stopped on the Antarctic-former Phoenix plate boundary at about 4 Ma as given by *Barker* [1982] or by 3.3 Ma per *Livermore et al.* [2000]. While the *Livermore et al.* [2000] time for cessation of spreading requires asymmetric spreading and a number of changes in the spreading rate, it does agree with the most reasonable match shown earlier by *Larter and Barker* [1991]. Powell Basin is shown as fully open at 30 Ma (Figure 15). In Figure 16 (28.4 Ma), the first dated magnetic anomalies in the western Scotia Sea indicate that there is little possibility that Drake Passage was still blocked in any way by continental fragments. The strongest evidence for early opening of a circum Antarctic seaway downstream of Drake Passage as given by a proto-ACC, is at Maud Rise. Consequently, it may be reasonable to assume that the first seaway was through Powell Basin, resulting in an ACC that ran south toward the margin of East Antarctica. There is additional evidence to the north for a later time, ~28.5 Ma [*Marino and Flores*, 2002] of a second event that affected ODP Site 1090 on Agulhas Ridge which supports the idea that a deep seaway through the Scotia Sea may have finally opened at the end of the early Oligocene or possibly later.

We have updated the Drake Passage opening with motions for some of the small, elevated, or possibly continental fragments in the Scotia Sea. There are now three data sets for magnetic anomalies in the western Scotia Sea: the works of *Barker and Burrell* [1977], *LaBrecque and Cande* [1985], and *Eagles et al.* [2005]. Many of the small block motions are unconstrained but are shown in plausible scenarios. The idea that part of the western Scotia Sea is older than 30 Ma is suggested by the major plate reconstructions shown by *Lawver and Gahagan* [1998], *Barker* [2001], the Ph.D. thesis of *Eagles* [2001], and *Eagles et al.* [2005, 2006]. Only if unequivocal, correlated magnetic anomalies are determined for all the basins of the Scotia Sea region or drilling to basement in a number of places in the Scotia Sea will the true time of the opening of a deep seaway through Drake Passage be made.

11. CONCLUSIONS

The time of the opening of a deepwater passage between Australia and East Antarctica is still problematic. If the extended, apparent continental margin off the George V and Oates coasts of East Antarctica was not in place before Cenozoic glaciation developed on East Antarctica, then there may have been a deepwater passage to the south of the STR as early as 40 Ma. If the South Tasman Saddle was produced as a result of active seafloor spreading in the L'Atalante Depression to the east and continental stretching between the Tasmania and the STR during the late Cretaceous, then there may have been a medium-depth seaway between Tasmania and the STR in the late Paleocene and into the Eocene. If the South Tasman Saddle was in fact closed to major water transport until the Oligocene or later, and the Oates Coast margin of East Antarctica was always the width it is presently, then initial opening between Australia and Antarctica may have been delayed until the Eocene-Oligocene boundary when the STR finally cleared the East Antarctic margin as discussed by *Lawver and Gahagan* [2003].

The opening of Drake Passage between South America and the Antarctic Peninsula is also problematic. Various continental fragments could have clogged a deepwater seaway until 30 Ma, but by 28 Ma, there is little possibility that continental fragments or other elevated blocks could have continued to block medium-depth water circulation. In addition, a medium-depth to deepwater passage may have been open through Powell Basin to the south by 33 Ma, but it is difficult to constrain the precise time of opening of Powell Basin. From the present-day map of the various fronts that make up what is referred to as the ACC (Figure 1), it is clear that the majority of the deep water flow exits either through the Shag Rocks Passage or to the west of it. If intensification of the ACC relied on opening of a deep Shag Rocks Passage, then that may not have occurred until middle Miocene based on the observed magnetic anomalies in the northeastern part of the west Scotia Sea. Middle Miocene as a time for intensification of the ACC is supported by the ∂O^{18} values presented by *Zachos et al.* [2008]. In conclusion, the best determination of the precise time of the opening of seaways will always be dramatic shifts in water conditions and in sedimentation style as recorded in benthic and planktonic diatoms and forams.

Acknowledgments. We appreciate the in-depth reviews of Rob Larter and Graham Eagles that helped this paper. This work has been supported by the PLATES project at the Institute for Geophysics, Jackson School of Geosciences, an industry-supported plate tectonics research effort. Fieldwork in the Scotia Sea region has been supported by grants from the National Science Foundation to Lawver and Dalziel, the latest grant was OPP-ANT-0636850. This is UTIG contribution number 2337.

REFERENCES

Barker, P. F. (1982), The Cenozoic subduction history of the Pacific margin of the Antarctic Peninsula, *J. Geol. Soc. London, 139,* 787–801.

Barker, P. F. (1995), Tectonic framework of the east Scotia Sea, in *Backarc Basins: Tectonics and Magmatism,* edited by B. Taylor, pp. 281–314, Plenum, New York.

Barker, P. F. (2001), Scotia Sea regional tectonic evolution: Implications for mantle flow and palaeocirculation, *Earth Sci. Rev.*, *55*, 1–39.

Barker, P. F., and J. Burrell (1977), The opening of the Drake Passage, *Mar. Geol.*, *25*, 15–34.

Bauer, K., S. Neben, B. Schreckenberger, R. Emmermann, K. Hinz, N. Fechner, K. Gohl, A. Schulze, R. B. Trumbull, and K. Weber (2000), Deep structure of the Namibia continental margin as derived from integrated geophysical studies, *J. Geophys. Res.*, *105*, 25,829–25,853.

Berggren, W. A., D. V. Kent, C. C. Swisher III, and M. P. Aubry (1995), A revised Cenozoic geochronology and chronostratigraphy, in *Geochronology, Time Scales and Global Stratigraphic Correlation*, edited by W. A. Berggren et al., *Spec. Publ. SEPM Soc. Sediment. Geol.*, *54*, 129–212.

Bond, M., M. A. Reguero, S. E. Vizcaíno, and S. A. Marenssi (2006), A new 'South American ungulate' (Mammalia: Litopterna) from the Eocene of the Antarctic Peninsula, in *Cretaceous-Tertiary High-Latitude Palaeoenvironments: James Ross Basin, Antarctica*, edited by J. E. Francis, D. Pirrie, and J. A. Crame, *Geol. Soc. Spec. Publ.*, *258*, 163–176.

Boreham, C. J., J. E. Blevin, I. Duddy, J. Newman, K. Liiu, H. Middleton, M. K. Macphail, and A. C. Cook (2002), Exploring the potential for oil generation, migration and accumulation in Cape Sorell-1, Sorell Basin, offshore west Tasmania, *APPEA J.*, *42*, 405–435.

Brown, B., C. Gaina, and R. D. Müller (2006), Circum-Antarctic palaeobathymetry: Illustrated examples from Cenozoic to recent times, *Palaeogeogr. Palaeoclimatol. Palaeoecol.*, *231*, 158–168.

Bullard, E., J. E. Everett, and A. G. Smith (1965), The fit of the continents around the Atlantic, *Philos. Trans. R. Soc. London, Ser. A*, *285*, 41–51.

Cande, S. C., and D. V. Kent (1995), Revised calibration of the geomagnetic polarity timescale for the Late Cretaceous and Cenozoic, *J. Geophys. Res.*, *100*, 6093–6095.

Cande, S. C., and J. C. Mutter (1982), A revised identification of the oldest sea-floor spreading anomalies between Australia and Antarctica, *Earth Planet. Sci. Lett.*, *58*, 151–160.

Cande, S. C., and J. M. Stock (2004a), Cenozoic reconstructions of the Australian-New Zealand-South Pacific sector of Antarctica, in *The Cenozoic Southern Ocean: Tectonics, Sedimentation, and Climate Change Between Australia and Antarctica*, Geophys. Monogr. Ser., vol. 151, edited by N. F. Exon, J. P. Kennett, and M. J. Malone, pp. 5–17, AGU, Washington, D. C.

Cande, S. C., and J. M. Stock (2004b), Pacific-Antarctic-Australia motion and the formation of the Macquarie plate, *Geophys. J. Int.*, *157*, 399–414.

Cande, S. C., J. M. Stock, R. D. Müller, and T. Ishihara (2000), Cenozoic motion between east and west Antarctica, *Nature*, *404*, 145–150.

Casadio, S., C. Nelson, P. Taylor, M. Griffen, and D. Gordon (2010), West Antarctic Rift system: A possible New Zealand-Patagonia Oligocene paleobiogeographic link, *Ameghiniana (Rev. Assoc. Paleontol. Argentina)*, *47*(1), 129–132.

Cooper, A., H. Stagg, and E. Geist (1991), Seismic stratigraphy and structure of Prydz Bay, Antarctica: Implications from Leg 119 drilling, *Proc. Ocean Drill. Program Sci. Results*, *119*, 5–26.

Crawford, A. J., R. Lanyon, M. Elmes, and S. Eggins (1997), Geochemistry and significance of basaltic rocks dredged from the south Tasman Rise and adjacent seamounts, *Aust. J. Earth Sci.*, *44*, 621–632.

Cunningham, S. A., S. G. Alderson, B. A. King, and M. A. Brandon (2003), Transport and variability of the Antarctic Circumpolar Current in Drake Passage, *J. Geophys. Res.*, *108*(C5), 8084, doi:10.1029/2001JC001147.

Cunningham, W. D., I. W. D. Dalziel, T.-Y. Lee, and L. A. Lawver (1995), Southernmost South America-Antarctic Peninsula relative plate motions since 84 Ma: Implications for the tectonic evolution of the Scotia Arc region, *J. Geophys. Res.*, *100*(B5), 8257–8266.

Dalziel, I. W. D., and L. A. Lawver (2001), The lithospheric setting of the West Antarctic Icesheet, in *The West Antarctic Ice Sheet: Behavior and Environment*, Antarct. Res. Ser., vol. 77, edited by R. B. Alley and R. M. Bindschadler, pp. 29–44, AGU, Washington, D. C.

Dalziel, I. W. D., R. H. Dott Jr., R. D. Winn Jr., and R. L. Bruhn (1975), Tectonic relations of South Georgia Island to the southernmost Andes, *Geol. Soc. Am. Bull.*, *86*, 1034–1040.

Duncan, R. A., and I. MacDougall (1989), Volcanic time-space relationships, in *Intraplate Volcanism in Eastern Australia and New Zealand*, edited by R. W. Johnson, pp. 43–54, Cambridge Univ. Press, Cambridge, U. K.

Eagles, G. (2001), Modelling plate kinematics in the Scotia Sea, Ph.D. thesis, Univ. of Leeds, Leeds, U. K.

Eagles, G. (2010a), The age and origin of the central Scotia Sea, *Geophys. J. Int.*, *183*, 587–600.

Eagles, G. (2010b), South Georgia and Gondwana's Pacific Margin: Lost in translation?, *J. South Am. Earth Sci.*, *30*, 65–70.

Eagles, G., and R. A. Livermore (2002), Opening history of Powell Basin, Antarctic Peninsula, *Mar. Geol.*, *185*, 197–207.

Eagles, G., K. Gohl, and R. D. Larter (2004), High-resolution animated tectonic reconstruction of the South Pacific and West Antarctic Margin, *Geochem. Geophys. Geosyst.*, *5*, Q07002, doi:10.1029/2003GC000657.

Eagles, G., R. A. Livermore, J. D. Fairhead, and P. Morris (2005), Tectonic evolution of the west Scotia Sea, *J. Geophys. Res.*, *110*, B02401, doi:10.1029/2004JB003154.

Eagles, G., R. A. Livermore, and P. Morris (2006), Small basins in the Scotia Sea: The Eocene Drake Passage gateway, *Earth Planet. Sci. Lett.*, *242*, 343–353.

Exon, N. F., and A. J. Crawford (1997), Introduction, *Aust. J. Earth Sci.*, *44*, 539–541.

Exon, N. F., P. J. Hill, and J.-Y. Royer (1995), New maps of crust off Tasmania expand research possibilities, *Eos Trans. AGU*, *76*, 201.

Exon, N. F., A. M. G. Moore, and P. L. J. Hill (1997a), Geological framework of the south Tasman Rise, south of Tasmania and its sedimentary basins, *Aust. J. Earth Sci.*, *44*, 561–577.

Exon, N. F., R. F. Berry, A. J. Crawford, and P. J. Hill (1997b), Geological evolution of the East Tasman Plateau, a continental fragment southeast of Tasmania, *Aust. J. Earth Sci.*, *44*, 597–608.

Exon, N. F., J. P. Kennett, and M. J. Malone (2004), Leg 189 synthesis: Cretaceous-Holocene history of the Tasmanian Gateway [online], *Proc. Ocean Drill. Program Sci. Results*, *189*, 37 pp. [Available at http://www-odp.tamu.edu/publications/189_SR/VOLUME/SYNTH/SYNTH.PDF.]

Expedition 318 Scientists (2010), Wilkes Land Glacial History: Cenozoic East Antarctic Ice Sheet evolution from Wilkes Land margin sediments, *Integr. Ocean Drill. Program Prelim. Rep.*, *318*, doi:10.2204/iodp.pr.318.2010.

Finn, C., D. Moore, D. Damaske, and T. Mackey (1999), Aeromagnetic legacy of early paleozoic subduction along the Pacific margin of Gondwana, *Geology*, *27*, 1087–1090.

Foden, J. D., M. A. Elburg, S. P. Turner, M. Sandiford, J. O'Callaghan, and S. Mitchell (2002), Granite production in the Delamerian Orogen, South Australia, *J. Geol. Soc. London*, *154*, 557–575.

Foster, D. A., and A. J. W. Gleadow (1992), Reactivated tectonic boundaries and implications for the reconstruction of southeastern Australia and northern Victoria Land, Antarctica, *Geology*, *20*, 267–270.

Gahagan, L. M., et al. (1988), Tectonic fabric map of the ocean basins from satellite altimetry data, *Tectonophysics*, *155*, 1–26.

Gaina, C., D. R. Müller, J.-Y. Royer, J. Stock, J. Hardebeck, and P. Symonds (1998), The tectonic history of the Tasman Sea: A puzzle with 13 pieces, *J. Geophys. Res.*, *103*, 12,413–12,433.

Galindo-Zaldivar, J., A. Jabaloy, A. Maldonado, J. M. Martinez-Martinez, C. Sanz de Galdeano, L. Somoza, and E. Surinach (2000), Deep crustal structure of the area of intersection between the Shackleton fracture zone and the West Scotia Ridge (Drake Passage, Antarctica), *Tectonophysics*, *320*, 123–139.

Gee, J. S., and D. V. Kent (2007), Source of oceanic magnetic anomalies and the geomagnetic polarity timescale, in *Treatise on Geophysics*, vol. 5, *Geomagnetism*, pp. 455–507, Elsevier Sci., Amsterdam.

Ghiglione, M. C., D. Yagupsky, M. Ghidella, and V. A. Ramos (2009), Continental stretching preceding the opening of the Drake Passage: Evidence from Tierra del Fuego, *Geology*, *36*(8), 643–646, doi:10.1130/G24857A.1.

Gladczenko, T. P., K. Hinz, O. Eldholm, H. Meyer, S. Neben, and J. Skogseid (1997), South Atlantic volcanic margins, *J. Geol. Soc. London*, *154*, 465–470.

Godthelp, H., S. Wroe, and M. Archer (1999), A new marsupial from the early Eocene Tingamarra local fauna of Murgon, southeastern Queensland: A prototypical Australian marsupial?, *J. Mammal. Evol.*, *6*(3), 289–313.

Goin, F. J., R. Pascual, W. V. Koenigswald, M. O. Woodburne, J. A. Case, M. Reguero, and S. F. Vizcaíno (2006), First Gondwanatherian Mammal from Antarctica, in *Cretaceous-Tertiary High-Latitude Paleoenvironments, James Ross Basin, Antarctica*, edited by J. E. Francis, D. Pirrie, and J. A. Crame, *Geol. Soc. Spec. Publ.*, *258*, 135–144.

Goin, F. J., N. Zimicz, M. A. Reguero, S. N. Santillana, S. A. Marenssi, and J. J. Moly (2007), New marsupial (Mammalia) from the Eocene of Antarctica, and the origins and affinities of the microbiotheria, *Rev. Asoc. Geol. Argentina*, *62*(4), 597–603.

Goldstein, S. L., and S. R. Hemming (2003), Long-lived isotopic tracers in oceanography, paleoceanography, and ice-sheet dynamics, in *Treatise on Geochemistry*, vol. 6, in *The Oceans and Marine Geochemistry*, edited by H. Elderfield, pp. 453–489, Elsevier, Amsterdam.

Gordon, A. L. (1986), Interocean exchange of thermocline water, *J. Geophys. Res.*, *91*(C4), 5037–5047.

Gordon, A. L., and R. A. Fine (1996), Pathways of water between the Pacific and Indian oceans in the Indonesian seas, *Nature*, *379*, 146–149.

Gordon, A. L., C. F. Giulivi, and A. G. Ilahude (2003), Deep topographic barriers within the Indonesian seas, *Deep Sea Res., Part II*, *50*, 2205–2228.

Hill, I. A., and P. F. Barker (1980), Evidence for Miocene back-arc spreading in the central Scotia Sea, *Geophys. J. R. Astron. Soc.*, *63*, 427–440.

Hill, P. J., A. M. G. Moore, and N. F. Exon (2001), Sedimentary basins and structural framework of the South Tasman Rise and East Tasman Plateau, in *Eastern Australasian Basins Symposium, 2001; A Refocused Energy Perspective for the Future*, *Spec. Publ.*, vol. 1, edited by K. C. Hill and T. Bernecker, pp. 37–48, Pet. Explor. Soc. of Aust., Sydney, N. S. W., Australia.

Hinz, K., M. Hemmerich, U. Salge, and O. Eiken (1990), Structures in rift-basin sediments on the conjugate margins of western Tasmania, South Tasman Rise, and Ross Sea, Antarctica, in *Proceedings of the 1988 NATO Advanced Research Workshop on Geological History of the Polar Oceans: Arctic Versus Antarctic*, *NATO ASI Ser., Ser. C*, vol. 308, edited by U. Bleil, pp. 119–130, Springer, New York.

Kennett, J. P. (1977), Cenozoic evolution of Antarctic Glaciation, the circum-Antarctic Ocean, and their impact on global paleoceanography, *J. Geophys. Res.*, *82*, 3843–3859.

Kennett, J. P., and N. J. Shackleton (1976), Oxygen isotopic evidence for the development of the psychrosphere 38 Myr ago, *Nature*, *260*, 512–515.

Klepeis, K. A., and L. A. Lawver (1996), Tectonics of the Antarctic-Scotia plate boundary near Elephant and Clarence Islands, West Antarctica, *J. Geophys. Res.*, *101*, 20,211–20,231.

LaBrecque, J. L., and S. C. Cande (1985), Total intensity magnetic anomaly profiles, south, in *South Atlantic Ocean and Adjacent Antarctic Continental Margin*, edited by J. J. LaBrecque, Mar. Sci. Int., Woods Hole, Mass.

LaBrecque, J. L., and P. D. Rabinowitz (1977), Magnetic anomalies bordering the continental margin of Argentina, *Map 826*, Am. Assoc. of Pet. Geol., Tulsa, Okla.

Lanyon, R., R. Varne, and A. J. Crawford (1993), Tasmanian Tertiary basalts, the Balleny plume, and opening of the Tasman Sea (soutwest Pacific Ocean), *Geology*, *21*, 555–558.

Lawver, L. A., and L. M. Gahagan (1998), Opening of Drake Passage and its impact on Cenozoic ocean circulation, in *Tectonic Boundary Conditions for Climate Reconstructions*, *Oxford Monogr. Geol. Geophys.*, vol. 39, edited by T. J. Crowley and K. C. Burke, pp. 212–223, Oxford Univ. Press, Oxford, U. K.

Lawver, L. A., and L. M. Gahagan (2003), Evolution of Cenozoic seaways in the circum-Antarctic region, in *Antarctic Cenozoic Palaeoenvironments: Geologic Record and Models*, edited by F. Florindo, and A. Cooper, *Palaeogeogr. Palaeoclimatol. Palaeoecol.*, *198*, 11–37.

Lawver, L. A., L. M. Gahagan, and I. W. D. Dalziel (1998), A tight fit-early Mesozoic Gondwana, a plate reconstruction perspective, in *Origin and Evolution of Continents*, edited by Y. Motoyoshi and K. Shiraishi, *Mem. Natl. Inst. Polar Res. Spec. Issue (Jpn.)*, *53*, 214–229.

Ling, H. F., K. W. Burton, R. K. O'Nions, B. S. Kamber, F. von Blanckenburg, A. J. Gibb, and J. R. Hein (1997), Evolution of Nd and Pb isotopes in central Pacific seawater from ferro-manganese crusts, *Earth Planet. Sci. Lett.*, *146*, 1–12.

Livermore, R., et al. (2000), Autopsy on a dead spreading center: The Phoenix Ridge, Drake Passage, Antarctica, *Geology*, *28*, 607–610.

Livermore, R. A., G. Eagles, P. Morris, and A. Maldonado (2004), Shackleton Fracture Zone: No barrier to early circumpolar ocean circulation, *Geology*, *32*, 797–800.

Livermore, R. A., A. P. Nankivell, G. Eagles, and P. Morris (2005), Paleogene opening of Drake Passage, *Earth Planet. Sci. Lett.*, *236*, 459–470.

Lyle, M., S. Gibbs, T. C. Moore, and D. K. Rea (2007), Late Oligocene initiation of the Antarctic Circumpolar Current: Evidence from the South Pacific, *Geology*, *35*, 691–694, doi:10.1130/G23806A.1.

Lythe, M. B., D. G. Vaughan, and the BEDMAP Consortium (2000), BEDMAP- Bed topography of the Antarctic, *Misc 9*, scale 1:10,000,000, Br. Antarct. Surv., Cambridge, U. K.

Marino, M., and J.-A. Flores (2002), Middle Eocene to early Oligocene calcareous nannofossil stratigraphy at Leg 177 Site 1090, *Mar. Micropaleontol.*, *45*, 383–398.

Marks, K. M., J. M. Stock, and K. J. Quinn (1999), Evolution of the Australian-Antarctic discordance since Miocene time, *J. Geophys. Res.*, *104*, 4967–4981.

Müller, R. D., W. R. Roest, and J.-Y. Royer (1998), Asymmetric sea-floor spreading caused by ridge-plume interactions, *Nature*, *396*, 455–459.

Müller, R. D., J.-Y. Royer, S. C. Cande, W. R. Roest, and S. Maschenkov (1999), New constraints on the Late Cretaceous/Tertiary plate tectonic evolution of the Caribbean, in *Caribbean Basins*, *Sediment. Basins World*, vol. 4, edited by P. Mann, pp. 33–59, Elsevier, Amsterdam.

Mutter, J. C., K. A. Hegarty, S. C. Cande, and J. K. Weissel (1985), Breakup between Australia and Antarctica: A brief review in light of new data, *Tectonophysics*, *114*, 255–280.

Naveira Garabato, A. C., D. P. Stevens, and K. J. Heywood (2002), Modification and pathways of Southern Ocean deep waters in the Scotia Sea, *Deep Sea Res., Part I*, *49*, 681–705.

Neil, H. L., L. Carter, and M. Y. Morris (2004), Thermal isolation of Campbell Plateau, New Zealand, by the Antarctic Circumpolar Current over the past 130 kyr, *Paleoceanography*, *19*, PA4008, doi:10.1029/2003PA000975.

Nilsson, M. A., G. Churakov, M. Sommer, N. V. Tran, A. Zemann, J. Brosius, and J. Schmitz (2010), Tracking marsupial evolution using archaic genomic Retroposon insertions, *PLoS Biol.*, *8*(7), e1000436, doi:10.1371/journal.pbio.1000436.

Pascual, R., F. J. Goin, L. Balarino, and D. E. Udrizar Sauthier (2002), New data on the Paleocene *Monotrematum sudamericanum* and the convergent evolution of triangulate molars, *Acta Palaeontol. Pol.*, *47*, 487–492.

Pfuhl, H. A., and I. N. McCave (2005), Evidence for late Oligocene establishment of the Antarctic Circumpolar Current, *Earth Planet. Sci. Lett.*, *235*, 715–728, doi:10.1016/j.epsl.2005.04.025.

Reguero, M. A., and S. A. Marenssi (2010), Paleogene climatic and biotic events in the terrestrial record of the Antarctic Peninsula: An overview, in *The Paleontology of Gran Barranca: Evolution and Environmental Change Through the Middle Cenozoic of Patagonia*, edited by R. H. Madden et al., pp. 383–397, Cambridge Univ. Press, Cambridge, U. K.

Rignot, E., J. L. Bamber, M. R. Van den Broecke, C. Davis, Y. Li, W. J. Van de Berg, and E. Van Meijgaard (2008), Recent Antarctic ice mass loss from radar interferometry and regional climate modelling, *Nat. Geosci.*, *1*, 106–110, doi:10.1038/ngeo102.

Royer, J.-Y. (1992), The opening of the Indian Ocean since the Late Jurassic: An overview, in *Proceedings of the Indian Ocean First Seminar on Petroleum Exploration, Seychelles*, edited by P. S. Plummer, pp. 169–185, U. N. Dep. of Tech. Coop. Dev., New York.

Royer, J.-Y., and T. Chang (1991), Evidence for relative motions between the Indian and Australian plates during the last 20 Myr from plate tectonic reconstructions: Implications for the deformation of the Indo-Australian plate, *J. Geophys. Res.*, *96*, 11,779–11,802.

Royer, J.-Y., and N. Rollet (1997), Plate-tectonic setting of the Tasmanian region, *Aust. J. Earth Sci.*, *44*, 543–560.

Royer, J.-Y., and D. T. Sandwell (1989), Evolution of the eastern Indian Ocean since the Late Cretaceous: Constraints from Geosat altimetry, *J. Geophys. Res.*, *94*, 13,755–13,782.

Royer, J.-Y., P. Patriat, H. Bergh, and C. Scotese (1988), Evolution of the southwest Indian Ridge from the Late Cretaceous (anomaly 34) to the middle Eocene (anomaly 20), *Tectonophysics*, *155*, 235–260.

Sandwell, D. T., and W. H. F. Smith (1997), Marine gravity anomaly from Geosat and ERS-1 satellite altimetry, *J. Geophys. Res.*, *102*, 10,039–10,054.

Sandwell, D. T., and B. Zhang (1989), Global mesoscale variability from the Geosat exact repeat mission: Correlation with ocean depth, *J. Geophys. Res.*, *94*, 17,971–17,984.

Scher, H. D., and E. T. Martin (2004), Timing and climatic consequences of the opening of Drake Passage, *Science*, *312*, 228–230.

Smith, W. H. F., and D. T. Sandwell (1997), Global sea floor topography from satellite altimetry and ship depth soundings, *Science*, *277*, 1956–1962.

Stagg, H. M. J., J. B. Willcox, P. A. Symonds, G. W. O'Brien, J. B. Colwell, P. J. Hill, D.-S. Lee, A. M. G. Moore, and H. I. M. Struckmeyer (1999), Architecture and evolution of the Australian continental margin, *AGSO J. Aust. Geol. Geophys.*, *17*, 17–33.

Stickley, C. E., H. Brinkhuis, S. A. Schellenberg, A. Sluijs, U. Rohl, M. Fuller, M. Grauert, M. Huber, J. Warnaar, and G. L. Williams (2004), Timing and nature of the deepening of the Tasmanian Gateway, *Paleoceanography*, *19*, PA4027, doi:10.1029/2004PA001022.

Talwani, M., J. Mutter, R. Houtz, and M. König (1979), The crustal structure and evolution of the area underlying the Magnetic Quiet Zone on the margin south of Australia, in *Geologic and Geophysical Investigations of Continental Slopes and Rises*, edited by J. Watkins and J. Montadert, *AAPG Mem.*, *29*, 151–175.

Thomas, D. J., T. J. Bralower, and C. E. Jones (2003), Neodymium isotopic reconstruction of late Paleocene-early Eocene thermohaline circulation, *Earth Planet. Sci. Lett.*, *209*, 309–322.

Tikku, A. A., and S. C. Cande (2000), On the fit of Broken Ridge and Kerguelen plateau, *Earth Planet. Sci. Lett.*, *180*, 117–132.

Veevers, J. J. (1986), Breakup of Australia and Antarctica estimated as mid-Cretaceous (95+5 Ma) from magnetic and seismic data at the continental margin, *Earth Planet. Sci. Lett.*, *77*, 91–99.

Vogt, P. R., W.-Y. Jung, and J. Brozena (1998), Arctic margin gravity highs; deeper meaning for sediment depocenters?, *Mar. Geophys. Res.*, *20*, 459–477.

Wilcox, J. B., and H. M. J. Stagg (1990), Australia's southern margin: A product of oblique extension, *Tectonophysics*, *173*, 269–281.

Woodburne, M. O., and W. J. Zinsmeister (1984), The first land mammal from Antarctica and its biogeographic implications, *J. Paleontol.*, *58*, 913–948.

Zachos, J. C., M. Pagani, L. Sloan, E. Thomas, and K. Billups (2001), Trends, rhythms, and aberrations in global climate 65 Ma to present, *Science*, *292*, 686–693.

Zachos, J. C., G. R. Dickens, and R. E. Zeebe (2008), An early Cenozoic perspective on greenhouse warming and carbon-cycle dynamics, *Nature*, *451*, 279–283.

I. W. D. Dalziel, L. M. Gahagan, and L. A. Lawver, Institute for Geophysics, University of Texas at Austin, 10100 Burnet Rd.-R2200, Austin, TX 78758-4445, USA. (lawver@ig.utexas.edu)

Exhumational History of the Margins of Drake Passage From Thermochronology and Sediment Provenance

David L. Barbeau Jr.

Department of Earth and Ocean Sciences, University of South Carolina
Columbia, South Carolina, USA

The margins of Drake Passage contain a rich and important perspective on the opening history of the marine gateway that until recently has been largely underexploited. In this chapter, I summarize and compare the Cretaceous and Paleogene exhumational history of the Fuegian Andes and the northern Antarctic Peninsula from existing thermochronology and sediment provenance data. Fission track and (U-Th-Sm)/He thermochronology of apatite and zircon collected from rocks in the Antarctic Peninsula indicate rapid exhumation-related cooling in the Late Cretaceous, whereas similar data from the Fuegian Andes indicate rapid exhumation in the Eocene, when existing data indicate exhumational cooling in the Patagonian Andes was comparatively slow, in general. U-Pb detrital zircon geochronology of Paleocene-Oligocene(?) strata from the Larsen Basin of the northern Antarctic Peninsula is fundamentally different from that of equivalently aged strata in the eastern Magallanes Basin of Tierra del Fuego, requiring sedimentary separation of these two basins prior to the Paleocene. Although equivocal, together these data may support an onset of Drake Passage opening significantly prior to the oldest current estimates and partially culminating in the middle Eocene when a shortening gradient in the southern Andes may have enabled oroclinal bending in the southernmost Andes.

1. INTRODUCTION

The importance of the tectonic opening of Drake Passage for the Cenozoic cooling and glaciation of Antarctica is widely debated and incompletely constrained. Recent climate modeling and atmosphere-proxy studies suggest that drawdown of carbon dioxide [*DeConto and Pollard*, 2003; *Pagani et al.*, 2005] may have been the driver of earliest Oligocene (circa 34 Ma) glaciation of Antarctica. However, a sizable body of recent work [*Eagles et al.*, 2006; *Scher and Martin*,

Tectonic, Climatic, and Cryospheric Evolution of the Antarctic Peninsula
Special Publication 063
Copyright 2011 by the American Geophysical Union.
10.1029/2010SP000992

35

2006; *Livermore et al.*, 2007; *Ghiglione et al.*, 2008; *Barbeau et al.*, 2009; *Gombosi et al.*, 2009] supports the 30 year old hypothesis that ties the middle Cenozoic transition from greenhouse to icehouse states to changes in ocean circulation caused by opening of Drake Passage and the subsequent development of the Antarctic Circumpolar Current (ACC) [*Barker and Burrell*, 1977; *Kennett*, 1977].

Determining the evolution of Drake Passage opening is required to effectively evaluate the relative importance of thermal isolation of Antarctica by the ACC for these and other aspects of Antarctic climate evolution. Although several techniques and approaches have been brought to bear in constraining the geochronology of opening, challenges presented by the Scotia plate's small area, its complex mosaic of spreading centers and logistically challenging setting (Figure 1) have hampered a consensus for Drake Passage's age [e.g., *Lodolo et al.*, 2010]. Oceanic crust of the Scotia Sea has not been directly dated, but most paleomagnetic estimates for large-scale ocean crust production in Drake Passage are generally younger than circa 28 Ma [*Barker*, 2001; *Eagles et al.*, 2005], thereby postdating the earliest Oligocene Oi-1 glaciation of Antarctica [*Miller et al.*, 1987; *Zachos et al.*, 2001]. However, there is growing evidence of localized middle and late Eocene extensional subsidence and ocean crust production in small subbasins in the southern Scotia Sea [*Eagles et al.*, 2006] and southernmost South America [*Ghiglione et al.*, 2008], suggesting that the onset of ocean thoroughfare may well have preceded the dramatic growth and expansion of the Antarctic ice cap. Further support of pre-Oi-1 opening of Drake Passage comes from the neodymium isotopic composition of Eocene and Oligocene fossil fish teeth in the Atlantic sector of the Southern Ocean, which reveal a dramatic shift toward more juvenile values beginning around 41 Ma [*Scher and Martin*, 2006], suggesting an influx of significant volumes of Pacific-derived water to the southern Atlantic Ocean beginning in the middle Eocene.

Considering the aforementioned disparities in age estimates for Drake Passage opening along with the challenges in robustly dating small fragments of oceanic crust, it is desirable to employ as

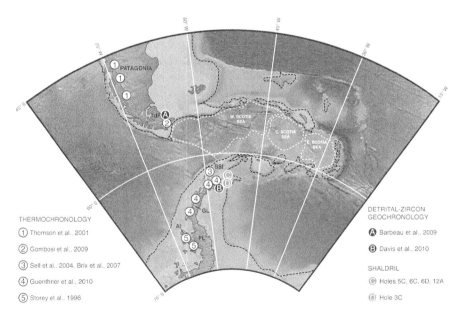

Figure 1. Map of southern South America, Drake Passage, and the Antarctic Peninsula depicting the geography and sample locations discussed in the text. Abbreviations are TdF, Tierra del Fuego; SSI, South Shetland Islands; GL, Graham Land; PL, Palmer Land; and AI, Alexander Island.

broad a range of tectonic proxies as possible in addressing the problem. In this chapter, I compile and interpret recent thermochronologic and sediment provenance data from the Antarctic Peninsula (AP) and Tierra del Fuego, which together record the exhumational history of the southern and northern margins of Drake Passage and serve as complementary tectonic proxies that may have implications for understanding the timing and nature of gateway opening and the subsequent tectonic and climatic evolution of the AP. Although the existing thermochronology and sediment provenance data sets are not yet sufficiently complete in their spatial and temporal distribution to fully resolve existing questions, their compilation should provide insight and constraints on existing and future interpretive tectonic models.

2. THERMOCHRONOLOGY

2.1. Thermochronology Overview

The field of thermochronology examines the timing and tempos of rock cooling [Reiners and Ehlers, 2005] and is frequently applied to constraining temperature changes that occur in association with tectonically mediated exhumation [Reiners and Brandon, 2006]. Thermochronometers function by producing radiogenic (daughter) isotopes or crystal damage caused by nuclear decay of radioactive (parent) isotopes in geologic materials and retaining these decay products within crystal lattices in concentrations proportional to the time elapsed since the sample cooled below a particular closure or "blocking" temperature. Moderate- and high-temperature thermochronometers (i.e., those with closure temperatures $\sim 350°C–900°C$) often record cooling ages that are congruent with or very close to crystallization ages. Low-temperature thermochronometers, however, preserve the thermal history of minerals at temperatures ($<250°C–350°C$) well below those minerals' solidi. As a result, low-temperature thermochronometers record many important geological processes that occur in the upper ~ 15 km of the lithosphere.

In geologic bodies distal to thermal perturbations such as igneous intrusions and hydrothermal systems, a low-temperature thermochronometer documents the timing of transit of the body upward through the isotherm equal to the system's closure temperature. By compiling data from (1) multiple samples acquired from a pseudovertical transect through different paleodepths and analyzed using a single thermochronometer [e.g., Fitzgerald et al., 1995] or (2) a single sample analyzed by multiple thermochronometers [e.g., Kohn et al., 1995; Kirstein et al., 2006], one can reconstruct the thermal history of a contiguous rock body. With some constraints on the four-dimensional thermal field of the investigated region during the time of cooling, these resulting time-temperature (t-T) pathways can allow quantitative interpretation of the studied region's exhumational history, from which the timing, rates, and magnitudes of rock uplift, a robust tectonic proxy, can often be plausibly inferred. In many recent studies of tectonically mediated exhumation, thermochronologists have widely employed fission track and (U-Th-Sm)/He analyses of the common crustal minerals apatite (calcium phosphate with hydroxyl, fluorine, chlorine, and bromine end-members) and zircon ($ZrSiO_4$).

In fission track thermochronology, the spontaneous fission of ^{238}U into two subequal product nuclides ranging in atomic number from 30 (zinc) to 65 (terbium) emits 200 MeV of energy [Faure, 1986] that is transferred to the product nuclides as kinetic energy, sending them in opposite directions and yielding a trail of radiation damage in the crystal lattice either ~ 11 μm long (in zircon) or ~ 16 μm long (in apatite). Above the closure temperatures for apatite (AFT) and zircon (ZFT) fission track thermochronometers, these radiation damage trails progressively anneal, largely by decreasing in length. However, once its closure temperature window is reached, the respective mineral retains its fission tracks, which when chemically etched are visible through high-magnification optical microscopy. The spatial density of these tracks with respect to the concentration of remaining ^{238}U (determined by mass spectrometry or induction of tracks from neutron bombardment of ^{235}U) allow the determination of the time elapsed since the grain last reached its closure temperature window [*Gleadow et al.*, 1986; *Green et al.*, 1989]. Moreover, the distribution of track lengths retained in a sample provides important constraints on the duration of time the sample spent in the closure temperature window (the partial annealing zone (PAZ) of *Gleadow and Fitzgerald* [1987]). For example, a narrow, symmetric, and unimodal distribution of track lengths close to the maximum possible track length (~11 μm for zircon and ~16 μm for apatite) indicates rapid passage of a sample through the PAZ, whereas a more broad distribution of track lengths indicates slow passage through the PAZ, and a multimodal distribution records a complicated history of with more than one passage into or through the PAZ [*Gleadow et al.*, 1986].

In contrast, (U-Th-Sm)/He thermochronology utilizes the production of alpha particles (radiogenic ^{4}He) from the decay of radioactive nuclides ^{238}U, ^{235}U, ^{232}Th, and ^{147}Sm. Because of the gaseous nature of radiogenic helium, it easily diffuses out of the crystal lattice above the containing mineral's closure temperature. Upon cooling below the closure temperatures for apatite (AHe) and zircon (ZHe) thermochronometers, the radiogenic helium is effectively retained because of lower diffusivity of crystal lattices at lower temperatures. Thus, its abundance relative to the remaining parent isotopes records the time elapsed since the grain last reached its closure temperature window [*Zeitler et al.*, 1987; *Wolf et al.*, 1996].

The approximate closure temperatures of these thermochronometers are largely dependent upon the mineral utilized, yet determination of a precise closure temperature can require specific knowledge (or reasonable assumption) of the geochemical composition, grain size, grain shape, cooling rate, and degree of radiation damage. As a rule of thumb, it may be useful to envision ~120°C and ~230°C as typical closure temperatures for AFT [*Ketcham et al.*, 1999] and ZFT [*Brandon et al.*, 1998] and ~70°C and ~180°C for AHe [*Farley*, 2000] and ZHe [*Reiners et al.*, 2004]. For a more robust treatment of effective closure temperature, the reader is referred to the work of *Reiners and Brandon* [2006] and references therein.

2.2. Thermochronology of Drake Passage Margins

In this chapter, I compile and interpret existing (U-Th-Sm)/He and fission-track cooling ages obtained from apatite and zircon collected from continental crust on the margins of Drake Passage in southern South America and the Antarctic Peninsula (Figure 1). Figure 2 depicts individual cooling ages obtained from thermochronometers collected from the central Patagonian Andes (Figure 2a) [*Thomson et al.*, 2001], the southern Patagonian Andes (Figure 2b) [*Thomson et al.*, 2001], the Fuegian Andes (Figure 2c) [*Gombosi et al.*, 2009], Graham Land and the South Shetland Islands (Figure 2d) [*Faundez et al.*, 2003; *Sell et al.*, 2004; *Brix et al.*, 2007; *Guenthner et al.*, 2010], and Alexander Island (Figure 2e) [*Storey et al.*, 1996]. In the case of individual samples from which multiple thermochronometers were obtained and analyzed, possible *t-T* pathways were constructed by linear interpolation between age constraints. When examining these possible pathways, it is important to consider the following caveats:

1. Not all reported data have sufficient contextual information to determine precise closure temperatures, as reflected in the qualitative vertical axes depicted in Figure 2.

2. Despite their parsimony, each interpolation represents but one possible *t-T* pathway. In reality, it is likely that the *t-T* path between any two constrained points was more complicated, as reflected in Monte Carlo modeling of multithermochronometer data that incorporates errors, a priori constraints, and other data such as track length distributions [*Ketcham*, 2005], none of

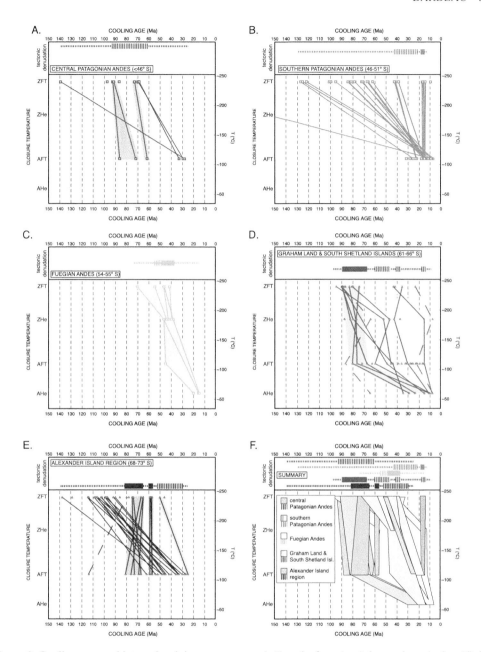

Figure 2. Cooling ages and interpolated time-temperature (*t-T*) paths from (a–c) the southern Andes, (d) the northern Antarctic Peninsula region, and (e) the southern Antarctic Peninsula region. (f) A summary of all regions for comparative purposes. Individual samples analyzed by more than one thermochronometer are interpolated to form one possible *t-T* path; see text for discussion and caveats. Periods of rapid cooling are indicated by colored polygons on graphs and by thick lines on summaries at the top of each plot. Dashed thick lines indicate rapid cooling in a relatively small subset of samples. See text for acronyms. Note that the closure temperature for each cooling age is approximate, and closure temperatures for each thermochronometer vary from sample to sample.

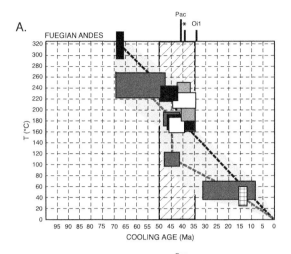

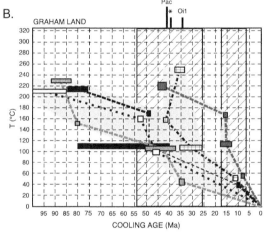

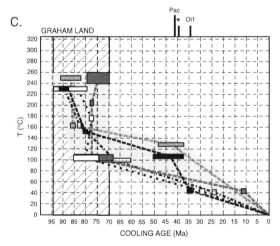

Figure 3

which are employed in the Figure 2 compilation. However, in the cases in which two thermo-chronometers from a single sample report relatively similar ages, some confidence can be assigned to that segment of the *t-T* pathway. For samples whose successive thermochronometers have notably different cooling ages, it is unlikely that linear interpolation fully records that segment's thermal history.

3. For simplicity in depiction, central age values are presented without any indication of the magnitudes of error or uncertainty. For reference, in the presented data, the average 1σ uncertainties are 7.8% for AHe ($n = 13$), 11.8% for AFT ($n = 94$), 3.6% for ZHe ($n = 18$), and 6.7% for ZFT ($n = 96$). There are no major differences in uncertainty in the different geographic regions, with the average uncertainty for each region ranging between 7% and 11%.

4. Although all thermochronology samples compiled herein come from hinterland crystalline rocks that were exhumed in association with Mesozoic-Cenozoic tectonics, they come from different tectonic domains, which may complicate simple comparisons. A large majority of samples from the Patagonian Andes and the Antarctic Peninsula come from large, felsic to intermediate Mesozoic plutons associated with subduction magmatism or the breakup of Gondwana; a minority come from metamorphic country rock that hosts the plutons. In contrast, the samples from the Fuegian Andes mostly come from small plutons preserved within deeply exhumed, large crystalline thrust sheets that are significantly inboard of the belt of plutonic rocks associated with subduction magmatism.

Keeping these caveats in mind, steep slopes in *t-T* pathways indicate relatively rapid cooling as is enabled by exhumation instigated by rapid rock uplift. Shallow negative slopes indicate slow cooling as could be achieved by postmagmatic dissipation of heat or slow exhumation caused by anorogenic erosion.

Fission track analysis of the central and southern Patagonian Andes [*Thomson et al.*, 2001] generally reveal widespread, moderate cooling rates throughout the Cretaceous and Paleogene Periods, with possible episodes of rapid cooling in the Late Cretaceous, late Paleogene, and middle Miocene (Figures 2a and 4b). In contrast, integrated fission track and (U-Th-Sm)/He dating of hinterland terranes from the Fuegian Andes [*Gombosi et al.*, 2009] reveal a widespread period of rapid cooling in the Eocene, with slow cooling leading up to and following the cooling event (Figure 2c).

Separate and integrated fission track and (U-Th-Sm)/He data from apatite and zircon collected from the South Shetland Islands [*Faundez et al.*, 2003; *Sell et al.*, 2004; *Brix et al.*, 2007] and Graham Land [*Faundez et al.*, 2003; *Sell et al.*, 2004; *Brix et al.*, 2007; *Guenthner et al.*, 2010] in the northern Antarctic Peninsula reveal dominant rapid cooling in the Late Cretaceous, with less well-constrained, but rapid cooling in the Paleogene and middle Miocene (Figure 2d). Separate and integrated fission-track data from apatite and zircon collected from the Alexander Island region [*Storey et al.*, 1996] adjacent to the southern Antarctic Peninsula (Figure 1) generally

Figure 3. (opposite) Fission track and (U-Th-Sm)/He cooling ages with 1σ errors, constrained blocking temperatures, and interpolated time-temperature paths for individual samples (indicated by pattern) collected from apatite and zircon aliquots from the Fuegian Andes of southern South America [*Gombosi et al.*, 2009] and Graham Land of the northern Antarctic Peninsula [*Guenthner et al.*, 2010]. Note the temporal relationships between the penetration of Pacific-derived ocean water through Drake Passage (Pac [*Scher and Martin*, 2006]), the rapid and dramatic sediment provenance change in the Magallanes foreland basin (asterisk [*Barbeau et al.*, 2009]), the Oi-1 expansion of continental ice on Antarctica (Oi1 [*Miller et al.*, 1987; *Zachos et al.*, 2001]), and rapid exhumation in the two regions (diagonally cross-hatched boxes). See text for discussion. See original contributions for information pertaining to specific samples.

record moderate rates of cooling throughout the Cretaceous, with some samples recording a narrow range of very rapid cooling in the Late Cretaceous (Figure 2e).

Figure 2f summarizes the well-defined and moderately defined periods of rapid and moderate low-temperature cooling on the northern and southern margins of Drake Passage. Well-defined Late Cretaceous rapid cooling is preserved in both the northern and southern Antarctic Peninsula region, whereas equivalently constrained cooling occurred in the Eocene in the Fuegian Andes. In contrast, in the Patagonian Andes, cooling generally appears to have occurred at slower rates, with the exception of periods of localized, rapid cooling in the late Paleogene and middle Miocene as recorded by the fission track method applied to cogenetic apatite and zircon [*Thomson et al.*, 2001].

The comparatively recent renaissance of (U-Th-Sm)/He thermochronology [*Reiners*, 2002] has enabled thermochronologists to acquire much higher-resolution low-temperature thermal histories than those previously afforded by fission track thermochronology alone. Figure 3 depicts all data acquired to date from individual thermochronology samples collected from the study region that were analyzed by both methods [*Gombosi et al.*, 2009; *Guenthner et al.*, 2010]. In contrast to Figure 2, these data include closure temperature estimates and uncertainties for both cooling age and closure temperature, as well as *t-T* pathways assigned by simple interpolation between central ages. The reader is referred to the original manuscripts for Monte Carlo simulations that better describe the range of likely *t-T* pathways. This analysis reveals the well-constrained rapid middle and late Eocene exhumation of the Fuegian Andes (Figure 3a) [*Gombosi et al.*, 2009], in contrast to moderate Paleogene (Figure 3b) and rapid Late Cretaceous (Figure 3c) cooling recorded in Graham Land in the northern Antarctic Peninsula [*Guenthner et al.*, 2010].

3. SEDIMENT PROVENANCE

3.1. Sediment Provenance Overview

The composition of detrital sediments preserved in sedimentary basins adjacent to active tectonic settings can provide valuable insight into the history of exhumation and the kinematic evolution of tectonically active source regions. In turn, these data can assess and often complement tectonic interpretations based on thermochronology and other tectonic proxies. Whereas a wide range of petrologic and geochemical fingerprinting techniques are employed in such sediment provenance analyses [e.g., *Dickinson and Suczek*, 1979; *McLennan et al.*, 1993; *Gehrels et al.*, 1995; *Goldstein et al.*, 1997; *Rahl et al.*, 2003; *Ross et al.*, 2005], zircon's refractory nature, its ubiquity in continental crustal rocks, and the high closure temperature of its U-Pb radiogenic isotope system have made it a popular and useful tool for accurately constraining and tracking the provenance of sediment delivered to a sedimentary basin [e.g., *Fedo et al.*, 2003]. Despite these benefits, sediment provenance analysis using U-Pb detrital zircon geochronology is not without challenges; successful application requires competent sediment-transport mechanisms, comparable zircon fertility and sufficient geochronologic diversity in candidate source regions, and minimization of fractionation biases.

3.2. Detrital Zircon Geochronology of the Margins of Drake Passage

In the remainder of this chapter, I present comparisons of the U-Pb geochronology of detrital zircons collected from Paleocene to Oligocene(?) sandstones of the Larsen Basin of the northern Antarctic Peninsula [*Davis et al.*, 2010] with those reported from similarly aged sandstones deposited in the eastern Magallanes Basin of southernmost South America [*Barbeau et al.*, 2009]. While plate reconstruction models for the region vary, southern South America and the northern Antarctic Peninsula were adjacent prior to the final breakup of Gondwana, with a restored distance between the sampling locations of these two suites (Figure 1) as little as 300 km in the Late Cretaceous, presumably prior to Drake Passage opening [e.g., *Lawver and Gahagan*, 2003; *Eagles et al.*, 2009; *Hayes et al.*, 2009].

In light of this proximity, similarities and differences in the sediment provenance of equivalently aged strata could provide useful constraints on the timing of continental separation. Statistical analysis [*Barbeau et al.*, 2010] of the detrital zircon age spectra from upper Paleozoic and Mesozoic sediments of southern South America [e.g., *Hervé et al.*, 2003] and the northern Antarctic Peninsula [*Barbeau et al.*, 2010] indicate a common parent population of sediments deposited on the two continents and provide the youngest minimum age constraint on sedimentary connections between the two continents.

Although magnetic anomaly interpretations of the oldest ocean floor in the western sector of the Southern Ocean [*König and Jokat*, 2006] indicate earliest Cretaceous ages (anomaly M17, circa 144 Ma), there is some general agreement that South America and the Antarctic Peninsula did not fully separate until at least the Eocene and perhaps not until much later (see Introduction). If so, the provenance of sediment deposited in the Magallanes and Larsen basins prior to the Eocene could be expected to be similar throughout the Mesozoic through to the early Paleogene. However, as depicted in Figure 4, comparison of composite U-Pb detrital zircon age spectra from Paleocene-Oligocene(?) strata in the two basins reveals fundamental differences that indicate derivation of sediment from separate source regions [*Davis et al.*, 2010]. A dominant

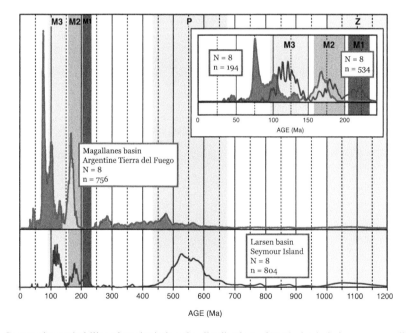

Figure 4. Composite probability plots depicting the distribution of U-Pb detrital zircon ages collected from broadly age-equivalent Paleocene-Oligocene(?) sandstones of the Magallanes and Larsen basins [*Barbeau et al.*, 2009; *Davis et al.*, 2010]. Lettered, colored boxes depict the major age populations from Seymour Island for comparison to the major populations of the eastern Magallanes Basin (colored probability plot infills). Abbreviations are Z, Precambrian age population present in Larsen Basin detrital zircons; P, Neoproterozoic-Paleozoic age population present in Larsen Basin detrital zircons; and M1, M2, and M3, successively younger Mesozoic-age subpopulations present in Larsen Basin detrital zircons.

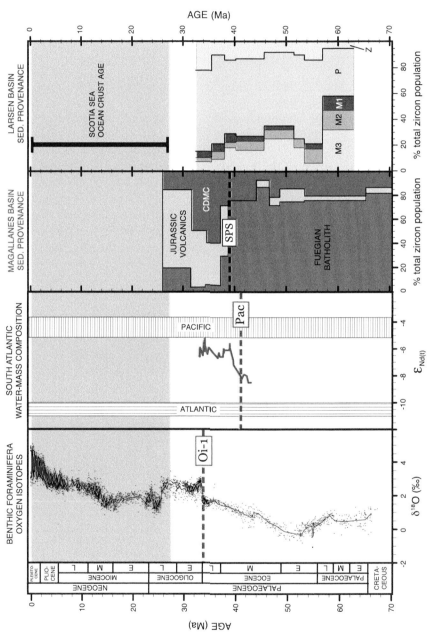

Figure 5. Comparison of the oxygen isotope composition of the global ocean [*Zachos et al.*, 2001], the Nd isotope geochemistry of the Atlantic sector of the Southern Ocean [*Scher and Martin*, 2006], and the detrital zircon sediment provenance of Late Cretaceous and Paleogene strata of the eastern Magallanes [*Barbeau et al.*, 2009] and Larsen basins [*Davis et al.*, 2010]. Abbreviations are Oi-1, rapid increase in seawater $\delta^{18}O$ associated with the expansion of the Antarctic ice cap at the Eocene-Oligocene transition; Pac, onset of the penetration of Pacific-derived ocean water through Drake Passage; and SPS, the rapid and dramatic sediment provenance change in the eastern Magallanes foreland basin. Colored polygons in the sediment provenance columns correspond to the age populations indicated in Figure 4.

Neoproterozoic and early Paleozoic zircon age population in Larsen Basin strata ("P" in Figure 4) is noticeably scant in the eastern Magallanes Basin, whereas the abundant Late Cretaceous and Paleogene zircons in the Magallanes Basin are all but entirely absent from the Larsen Basin. Although both basins contain Mesoproterozoic ("Z" in Figure 4), Jurassic ("M2"), and Early Cretaceous ("M3") age populations, the distribution of the Jurassic and Early Cretaceous populations are significantly different. A small Triassic age population ("M1") preserved in the Larsen Basin is absent from the eastern Magallanes Basin.

Although these results could be explained by a drainage divide that allowed sediment contributions to separate basins from partially distinct sources while the two continents were still attached, it is equally plausible that they indicate the opening of Drake Passage began prior to the Paleocene [*Davis et al.*, 2010]. Considering the statistical similarities between the composition of early Mesozoic strata on the two continents now separated by Drake Passage [*Barbeau et al.*, 2010] and the present lack of detrital zircon data from Larsen Basin Cretaceous strata, it is not yet possible to determine when prior to the Paleocene the provenances of the two continents diverged.

In addition to the fundamental differences between the composite detrital zircon age spectra of equivalent Paleogene strata in the eastern Magallanes and Larsen basins, the evolution of the sediment provenance of the two study regions contain insightful differences. A pronounced and rapid change in sediment provenance occurred in the middle Eocene in the Magallanes Basin (Figure 5) and provides independent support for rapid middle Paleogene shortening, rock uplift, and cooling in the Fuegian Andes [*Barbeau et al.*, 2009], as preserved in the orogen's thermochronology [*Gombosi et al.*, 2009]. In comparison, the detrital zircon composition of Paleogene strata in the Larsen Basin is remarkably uniform (Figure 5), consistent with the scant evidence for rapid Paleogene exhumation in the Antarctic Peninsula (Figures 2 and 3).

4. DISCUSSION

Although the thermochronology and sediment provenance data summarized herein are equivocal, they provide new constraints on models for the kinematic history of the Scotia Sea and the margins of Drake Passage:

1. The most dominant period of rapid exhumation in the northern Antarctic Peninsula since circa 100 Ma appears to have occurred during the Late Cretaceous [*Guenthner et al.*, 2010], when by some measures the Fuegian Andes were comparatively quiescent. Although the initial shortening phase in the Fuegian Andes began prior to circa 86 Ma [*Klepeis et al.*, 2010], sediment reaching the eastern Magallanes Basin throughout the Late Cretaceous was derived almost entirely from the Fuegian magmatic arc and associated units [*Barbeau et al.*, 2009]. Thus, a significant topographic barrier, as exists today in the form of the Cordilleran Darwin, between the arc and the foreland basin, does not appear to have existed prior to the middle Paleogene, as further indicated by slow pre-Paleogene exhumation rates from low-temperature thermochronology [*Gombosi et al.*, 2009].

2. The composition of sediments deposited in the Magallanes Basin of Tierra del Fuego and the Larsen Basin of the northern Antarctic Peninsula during the Paleocene and Eocene require that the two basins were separate, either by a topographic barrier or by prior continental separation. Paleocurrent and provenance interpretations from the Paleogene successions of the eastern Magallanes foreland basin [see *Barbeau et al.*, 2009] and the Larsen Basin [see *Marenssi et al.*, 2002] support primarily transverse sediment dispersal patterns, further raising the possibility that the sedimentary separation was caused by local variations in drainage divides rather than tectonic separation. However, the significantly large older Paleozoic and Proterozoic detrital zircon age

populations in Paleocene and Eocene strata of the Larsen Basin (Figure 4) do not appear to have sufficient appropriately aged sources in the adjacent rocks of the Antarctic Peninsula [*Millar et al.*, 2002] from which paleocurrent interpretations suggest the sediments were dispersed [*Marenssi et al.*, 2002]. Thus, the sediment dispersal pathways of at least the Larsen Basin remain poorly constrained.

3. The most dominant period of rapid exhumation in the Fuegian Andes since circa 100 Ma occurred in the middle Eocene [*Barbeau et al.*, 2009; *Gombosi et al.*, 2009], when the northern Antarctic Peninsula and Patagonian Andes were largely quiescent, as primarily indicated by thermochronology (Figure 3) [*Thomson et al.*, 2001; *Guenthner et al.*, 2010], and in case of the latter, a thin Paleogene foreland basin succession [*Biddle et al.*, 1986].

In light of these observations, new possibilities relevant to the history of Drake Passage opening may need to be further considered:

First, the markedly distinct sediment compositions of the Paleocene-Eocene Magallanes and Larsen basins (Figure 4) [*Barbeau et al.*, 2009; *Davis et al.*, 2010], the rapid exhumation in the northern Antarctic Peninsula (Figures 2d and 2f) [*Faundez et al.*, 2003; *Sell et al.*, 2004; *Brix et al.*, 2007; *Guenthner et al.*, 2010], and E-W oriented Cretaceous magnetic anomalies in the southern Weddell Sea (Figure 1) [*König and Jokat*, 2006] are all tectonic indicators that suggest Drake Passage may have begun opening as early as the Cretaceous, significantly prior to the oldest conventional estimates.

Second, the co-occurrence of the arrival of Pacific-derived seawater in the Atlantic sector of the Southern Ocean (Figure 5) [*Scher and Martin*, 2006] with rapid exhumation of the Fuegian Andes as indicated by thermochronology (Figure 2c) [*Gombosi et al.*, 2009] and a dramatic sediment provenance shift preserved in the eastern Magallanes Basin (Figure 5) [*Barbeau et al.*, 2009] suggests a possible genetic relationship between the opening history of Drake Passage and the shortening history of the southernmost Andes. Considering the relative tectonic quiescence in the Patagonian Andes and northern Antarctic Peninsula during this time (e.g., Figure 2), these results support a Paleogene development of the Patagonian orocline whereby greater shortening of the overriding continental plate in the Fuegian Andes than in the Patagonian Andes or the northern Antarctic Peninsula would foster separation of the two continents through a north-to-south shortening gradient in the southern Andes. Despite the simplicity of this model, recent interpretations of vertical-axis rotation data from the paleomagnetism of Eocene strata in the Magallanes Basin challenge such interpretations, while noting the mechanical paradox of north-vergent shortening in the Fuegian Andes when Antarctica and South America were apparently not converging [*Maffione et al.*, 2009].

Acknowledgments. Funding for this research comes from NSF grant OPP-0732995 to D. Barbeau, H. Scher, and R. Thunell for the Plates and Gates initiative of the 2007–2009 International Polar Year. The ideas and interpretations presented herein have benefitted from discussions with and the efforts of numerous students and colleagues, including N. Swanson-Hysell, D. Gombosi, W. Guenthner, K. Zahid, K. Murray, J. Davis, S. Hemming, G. Gehrels, H. Scher, E. Olivero, P. Reiners, S. Thomson, G. Eagles, J. Anderson, and J. Garver. However, any misconceptions or weaknesses in interpretation contained herein are solely the responsibility of the author. Constructive criticism from G. Eagles, J. Anderson, and an anonymous reviewer greatly improved the manuscript.

REFERENCES

Barbeau, D. L., E. B. Olivero, N. L. Swanson-Hysell, K. M. Zahid, K. E. Murray, and G. E. Gehrels (2009), Detrital-zircon geochronology of the eastern Magallanes foreland basin: Implications for Eocene kinematics of the northern Scotia Arc and Drake Passage, *Earth Planet. Sci. Lett.*, *20*, 23–45.

Barbeau, D. L., J. T. Davis, K. E. Murray, V. Valencia, G. E. Gehrels, K. M. Zahid, and D. J. Gombosi (2010), Detrital-zircon geochronology of metasedimentary rocks of northwestern Graham Land, *Antarct. Sci.*, *22*, 65–78, doi:10.1017/S095410200999054X.

Barker, P. (2001), Scotia Sea regional tectonic evolution: Implications for mantle flow and palaeocirculation, *Earth Sci. Rev.*, *55*, 1–39.

Barker, P. F., and J. Burrell (1977), The opening of Drake Passage, *Mar. Geol.*, *25*, 15–34.

Brandon, M. T., M. K. Roden-Tice, and J. I. Garver (1998), Late Cenozoic exhumation of the Cascadia accretionary wedge in the Olympic Mountains, northwest Washington State, *Geol. Soc. Am. Bull.*, *110*, 985–1009.

Biddle, K. T., M. A. Uliana, J. R. Mitchum, M. G. Fitzgerald, and R. C. Wright (1986), The stratigraphic and structural evolution of the central and eastern Magallanes Basin, southern South America, in *Foreland Basins*, edited by P. A. Allen and P. Homewood, *Spec. Publ. Int. Assoc. Sedimentol.*, *8*, 41–61.

Brix, M. R., V. Faundez, F. Hervé, M. Solari, J. Fernandez, A. Carter, and B. Stöckhert (2007), Thermo-chronologic constraints on the tectonic evolution of the western Antarctic Peninsula in late Mesozoic and Cenozoic times, in Antarctica: A Keystone in a Changing World — Online Proceedings of the 10th International Symposium on Antarctic Earth Sciences, edited by A. K. Cooper, C. R. Raymond, and 10th ISAES Editorial Team, *U.S. Geol. Surv. Open File Rep., 2007-1047*, 5 pp., doi:10.3133/of2007-1047.srp101. (Available at http://pubs.usgs.gov/of/2007/1047/srp/srp101/index.html.)

Davis, J. D., D. L. Barbeau, K. M. Zahid, K. E. Murray, and G. E. Gehrels (2010), Comparative U-Pb detrital-zircon analysis of southern South America and the northern Antarctic Peninsula, *Geol. Soc. Am. Abstr. Programs*, *42*, 128.

DeConto, R. M., and D. Pollard (2003), Rapid Cenozoic glaciation of Antarctica induced by declining atmospheric CO_2, *Nature*, *421*, 245–249.

Dickinson, W. R., and C. A. Suczek (1979), Plate tectonics and sandstone compositions, *AAPG Bull.*, *63*, 2164–2182.

Eagles, G., R. A. Livermore, J. D. Fairhead, and P. Morris (2005), Tectonic evolution of the west Scotia Sea, *J. Geophys. Res.*, *110*, B02401, doi:10.1029/2004JB003154.

Eagles, G., R. Livermore, and P. Morris (2006), Small basins in the Scotia Sea: The Eocene Drake Passage gateway, *Earth Planet. Sci. Lett.*, *242*, 343–353.

Eagles, G., K. Gohl, and R. D. Larter (2009), Animated tectonic reconstruction of the Southern Pacific and alkaline volcanism at its convergent margins since Eocene times, *Tectonophysics*, *464*, 21–29, doi:10.1016/j.tecto.2007.10.005.

Farley, K. A. (2000), Helium diffusion from apatite: General behavior as illustrated by Durango fluorapatite, *J. Geophys. Res.*, *105*, 2903–2914.

Faúndez, V., M. R. Brix, F. Hervé, S. N. Thomson, B. Stöckhert, and W. Loske (2003), Fission track thermochronology of the western Antarctic Peninsula and South Shetland Islands: A progress report on new zircon data, paper presented at X Congreso Geológico Chileno, Soc. Geol. de Chile, Concepción, Chile.

Faure, G. (1986), *Principles of Isotope Geology*, 2nd ed., 557 pp., John Wiley, New York.

Fedo, C. M., K. N. Sircombe, and R. H. Rainbird (2003), Detrital zircon analysis of the sedimentary record, in *Zircon: Experiments, Isotopes, and Trace Element Investigations: Mineralogical Society of America*, edited by J. M. Hanchar and P. Hoskin, *Rev. Mineral.*, *53*, 277–303.

Fitzgerald, P. G., R. B. Sorkhabi, T. F. Redfield, and E. Stump (1995), Uplift and denudation of the central Alaska range: A case study in the use of apatite fission track thermochronology to determine absolute uplift parameters, *J. Geophys. Res.*, *100*, 20,175–20,191.

Gehrels, G. E., W. R. Dickinson, G. M. Ross, J. H. Stewart, and D. G. Howell (1995), Detrital zircon reference for Cambrian to Triassic miogeoclinal strata of western North America, *Geology*, *23*, 831–834.

Ghiglione, M. C., D. Yagupsky, M. Ghidella, and V. A. Ramos (2008), Continental stretching preceding the opening of the Drake Passage: Evidence from Tierra del Fuego, *Geology*, *36*, 643–646.

Gleadow, A. J. W., and P. G. Fitzgerald (1987), Tectonic history and structure of the Transantarctic Mountains: New evidence from fission track dating in the Dry valleys area of southern Victoria Land, *Earth Planet. Sci. Lett.*, *82*, 1–14.

Gleadow, A. J. W., I. R. Duddy, P. F. Green, and J. F. Lovering (1986), Confined fission track lengths in apatite: A diagnostic tool for thermal history analysis, *Contrib. Mineral. Petrol.*, *94*, 405–415.

Goldstein, S. L., N. T. Arndt, and R. F. Stallard (1997), The history of a continent from U-Pb ages of zircons from Orinoco River sand and Sm-Nd isotopes in Orinoco Basin river sediments, *Chem. Geol.*, *139*, 271–286.

Gombosi, D. G., D. L. Barbeau, and J. I. Garver (2009), New thermochronometric constraints on the rapid Paleogene uplift of the Cordillera Darwin complex and related thrust sheets in the Fuegian Andes, *Terra Nova*, *21*, 507–515.

Green, P. F., I. R. Duddy, G. M. Laslett, K. A. Hegarty, A. J. W. Gleadow, and J. F. Lovering (1989), Thermal annealing of fission tracks in apatite: 4. Quantitative modelling techniques and extension to geological timescales, *Chem. Geol.*, *79*, 155–182.

Guenthner, W. R., D. L. Barbeau, P. W. Reiners, and S. N. Thomson (2010), Slab window migration and terrane accretion preserved by low-temperature thermochronology of a magmatic arc, northern Antarctic Peninsula, *Geochem. Geophys. Geosyst.*, *11*, Q03001, doi:10.1029/2009GC002765.

Hayes, D. E., C. Zhang, and R. A. Weissel (2009), Modeling paleobathymetry in the Southern Ocean, *Eos Trans. AGU*, *90*, 165.

Hervé, F., C. M. Fanning, and R. J. Pankhurst (2003), Detrital zircon age patterns and provenance in the metamorphic complexes of southern Chile, *J. South Am. Earth Sci.*, *16*, 107–123.

Kennett, J. P. (1977), Cenozoic evolution of Antarctic glaciation, the circum-Antarctic Ocean, and their impact on global paleoceanography, *J. Geophys. Res.*, *82*, 3843–3860.

Ketcham, R. A. (2005), Forward and inverse modeling of low-temperature thermochronometry data, in *Low-Temperature Thermochronology: Techniques, Interpretations, and Applications*, edited by P. W. Reiners and T. A. Ehlers, *Rev. Mineral. Geochem.*, *58*, 275–314.

Ketcham, R. A., R. A. Donelick, and W. D. Carlson (1999), Variability of apatite fission-track annealing kinetics. III. Extrapolation to geological time scales, *Am. Mineral.*, *84*, 1235–1255.

Kirstein, L. A., H. Sinclair, F. M. Stuart, and K. Dobson (2006), Rapid early Miocene exhumation of the Ladakh batholith, western Himalaya, *Geology*, *34*, 1049–1052.

Klepeis, K., P. Betka, G. Clarke, M. Fanning, F. Hervé, L. Rojas, C. Mpodozis, and S. Thomson (2010), Continental underthrusting and obduction during the Cretaceous closure of the Rocas Verdes rift basin, Cordillera Darwin, Patagonian Andes, *Tectonics*, *29*, TC3014, doi:10.1029/2009TC002610.

Kohn, M. J., F. S. Spear, M. T. Harrison, and I. W. D. Dalziel (1995), Ar/Ar geochronology and P–T–t paths from the Cordillera Darwin metamorphic complex, Tierra del Fuego, Chile, *J. Metamorph. Geol.*, *13*, 251–270.

König, M., and W. Jokat (2006), The Mesozoic breakup of the Weddell Sea, *J. Geophys. Res.*, *111*, B12102, doi:10.1029/2005JB004035.

Lawver, L. A., and L. M. Gahagan (2003), Evolution of Cenozoic seaways in the circum-Antarctic region, *Palaeogeogr. Palaeoclimatol. Palaeoecol.*, *198*, 11–37.

Livermore, R., C.-D. Hillenbrand, M. Meredith, and G. Eagles (2007), Drake Passage and Cenozoic climate: An open and shut case?, *Geochem. Geophys. Geosyst.*, *8*, Q01005, doi:10.1029/2005GC001224.

Lodolo, E., D. Civile, A. Vuan, A. Tassone, and R. Geletti (2010), The Scotia–Antarctica plate boundary from 35°W to 45°W, *Earth Planet. Sci. Lett.*, *293*, 200–215.

Maffione, M., F. Speranza, C. Faccenna, and E. Rosello (2010), Paleomagnetic evidence for a pre-early Eocene (~50 Ma) bending of the Patagonian orocline (Tierra del Fuego, Argentina): Paleogeographic and tectonic implications, *Earth Planet. Sci. Lett.*, *289*, 273–286.

Marenssi, S. A., L. I. Net, and S. N. Santillana (2002), Provenance, environmental and paleogeographic controls on sandstone composition in an incised-valley system: The Eocene La Meseta Formation, Seymour Island, Antarctica, *Sediment. Geol.*, *150*, 301–321.

McLennan, S. M., S. Hemming, D. K. McDaniel, and G. N. Hanson (1993), Geochemical approaches to sedimentation, provenance, and tectonics, in *Processes Controlling the Composition of Clastic Sediments*, edited by M. J. Johnson and A. Basu, *Spec. Pap. Geol. Soc. Am.*, *284*, 21–40.

Millar, I. L., R. J. Pankhurst, and C. M. Fanning (2002), Basement chronology of the Antarctic Peninsula: Recurrent magmatism and anatexis in the Palaeozoic Gondwana Margin, *J. Geol. Soc. London*, *159*, 145–157.

Miller, K. G., R. G. Fairbanks, and G. S. Mountain (1987), Tertiary oxygen isotope synthesis, sea level history, and continental margin erosion, *Paleoceanography*, *2*, 1–19.

Pagani, M., J. C. Zachos, K. H. Freeman, B. Tipple, and S. Bohaty (2005), Marked decline in atmospheric carbon dioxide concentrations during the Paleogene, *Science*, *309*, 600–603, doi:10.1126/science.1110063.

Rahl, J., P. W. Reiners, I. H. Campbell, S. Nicolescu, and C. M. Allen (2003), Combined single grain (U-Th)/He and U/Pb dating of detrital zircons from the Navajo Sandstone, Utah, *Geology*, *31*, 761–764.

Reiners, P. W. (2002), (U-Th)/He chronometry experiences a renaissance, *Eos Trans. AGU*, *83*(3), 21, doi:10.1029/2002EO000012.

Reiners, P. W., and M. T. Brandon (2006), Using thermochronology to understand orogenic erosion, *Annu. Rev. Earth Planet. Sci.*, *34*, 419–466.

Reiners, P. W., and T. A. Ehlers (Eds.) (2005), *Low-Temperature Thermochronology: Techniques, Interpretations, Applications*, *Rev. Mineral. Geochem.* *58*, 622 pp.

Reiners, P. W., T. L. Spell, S. Nicolescu, and K. A. Zanetti (2004), Zircon (U-Th)/He thermochronometry: He diffusion and comparisons with $^{40}Ar/^{39}Ar$ dating, *Geochim. Cosmochim. Acta*, *68*, 1857–1887.

Ross, G. M., P. J. Patchett, M. Hamilton, L. Heaman, P. G. DeCelles, E. Rosenberg, and M. K. Giovanni (2005), Evolution of the Cordilleran orogen (southwestern Alberta, Canada) inferred from detrital mineral geochronology, geochemistry, and Nd isotopes in the foreland basin, *Geol. Soc. Am. Bull.*, *117*, 747–763, doi:10.1130/B25564.1.

Scher, H. D., and E. E. Martin (2006), Timing and climatic consequences of the opening of Drake Passage, *Science*, *312*, 428–430.

Sell, I., G. Poupeau, J. M. González-Casado, and J. López-Martínez (2004), Fission track thermochronological study of King George and Livingston islands, South Shetland Islands (Western Antarctica), *Antarct. Sci.*, *16*, 191–197.

Storey, B. C., R. W. Brown, A. Carter, P. A. Doubleday, A. J. Hurford, D. I. M. Macdonald, and P. A. R. Nell (1996), Fission-track evidence for the thermotectonic evolution of a Mesozoic-Cenozoic fore-arc, Antarctica, *J. Geol. Soc. London*, *153*, 65–82.

Thomson, S. N., F. Hervé, and B. Stöckhert (2001), The Mesozoic–Cenozoic denudation history of the Pata-gonian Andes (southern Chile) and its correlation to different subduction processes, *Tectonics*, *20*, 693–711.

Wolf, R. A., K. A. Farley, and L. T. Silver (1996), Helium diffusion and low-temperature thermochronometry of apatite, *Geochim. Cosmochim. Acta*, *60*, 4231–4240.

Zachos, J., M. Pagani, L. Sloan, E. Thomas, and K. Billups (2001), Trends, rhythms, and aberrations in global climate 65 Ma to present, *Science*, *292*, 686–693.

Zeitler, P. K., A. L. Herczeg, I. McDougall, and M. Honda (1987), U-Th-He dating of apatite: A potential thermochronometer, *Geochim. Cosmochim. Acta*, *51*, 2865–2868.

D. L. Barbeau Jr., Department of Earth and Ocean Sciences, University of South Carolina, Columbia, SC 29208, USA. (dbarbeau@geol.sc.edu)

Seismic Stratigraphy of the Joinville Plateau: Implications for Regional Climate Evolution

R. Tyler Smith and John B. Anderson

Department of Earth Sciences, Rice University, Houston, Texas, USA

Seismic records from the southern margin of the Joinville Plateau show a sediment wedge with no apparent hiatuses or large unconformities throughout the late Oligocene to the early Pliocene section. Sedimentation rates calculated using both drill core and regional seismic lines show fairly constant rates of sedimentation, with infrequent and isolated episodes of slumping and turbidity current activity. Contour currents, associated with the Weddell Gyre, and glacimarine sedimentary processes have dominated sedimentation on the Joinville slope since at least the late Oligocene. Seismic and core data show no evidence of the Antarctic Peninsula Ice Sheet having grounded on the plateau until the early Pliocene. This is consistent with a phased expansion of the ice sheet from south to north across the northern peninsula.

1. INTRODUCTION

Seismic data collected as part of SHALDRIL II identified a sediment wedge located on the southern margin of the Joinville Plateau, located at the northern tip of the Antarctic Peninsula (Figure 1), as an ideal drilling location because of the wedge geometry and relative accessibility in ice-free waters. The wedge onlaps acoustic basement at an oblique angle. This results in increasingly older stratigraphic units being situated at or near the seafloor along the platform margin (Figure 2). The stratigraphic section is relatively complete, with key stratigraphic intervals sampled during the cruise including the late Oligocene, middle Miocene, early Pliocene, late Pliocene, and Pleistocene [see *Bohaty et al.*, this volume]. These drill cores provide a record of paleoclimatic, paleoenvironmental and paleoceanographic changes that complement the ice sheet history that is derived from onshore studies [*Smellie et al.*, 2006, 2008; *Hambrey et al.*, 2010] and from seismic and drill core results from the James Ross Basin [*Smith and Anderson*, 2010], situated just southwest of the study area (Figure 1). The intent of this chapter is to provide a sequence stratigraphic framework for the southern flank of Joinville Plateau and interpretation of depositional processes based on seismic facies analysis.

Tectonic, Climatic, and Cryospheric Evolution of the Antarctic Peninsula
Special Publication 063
10.1029/2010SP000980

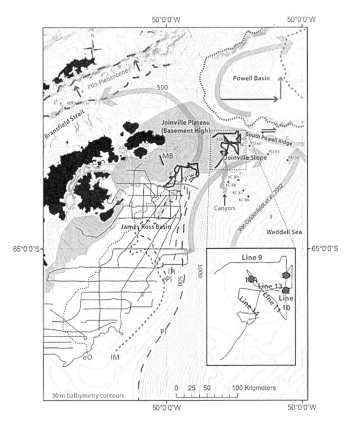

Figure 1. Regional geographic and bathymetric map showing locations of NBP0602A seismic lines (bold solid lines) and SHALDRIL II sites (large dots) used for this study. Fine lines are data used by *Smith and Anderson* [2010] in their seismic stratigraphic investigation of the James Ross Basin. Also shown are locations of Kasten cores (small dots designated with KC) and piston cores (small dots designated with PD) studied by *Gilbert et al.* [1998]. The large gray arrows indicate ocean current directions [*Von Gyldenfeldt et al.*, 2002]. The dotted and dashed lines show changes in the continental shelf break through time as follows: early Oligocene (eO), late Miocene (lM), early Pliocene (eP), late Pliocene (lPl), and Pleistocene (Pl) [*Smith and Anderson*, 2010]. To the north of the Joinville Plateau, the rifting away of the South Orkney Plateau and the opening of the Bransfield Strait are shown with dotted and dashed lines and arrows indicating directions of movement. The location of a prominent canyon/channel complex discussed in the text is also shown. MB indicates location of multibeam profile shown in Figure 6.

2. BACKGROUND

The Joinville Plateau is located east of Joinville Island and is the northernmost extension of the Antarctic Peninsula (Figure 1). It is bounded to the northwest by Bransfield Strait, to the northeast by Powell Basin and to the south by the Weddell Sea. The region is part of a back-arc basin that formed in response to subduction of the Phoenix and Aluk plates on the Pacific side of the Antarctic Peninsula during the Cretaceous [*Elliot*, 1988]. The Powell Basin was formed mainly during the Oligocene as the South Orkney Plateau rifted away from the Antarctic Peninsula [*Maldonado et al.*, 1998; *Eagles and Livermore*, 2002; *Lawver and Gahagan*, this volume]. The study area is situated along the southwestern flank of an E-W elongated morphological high, known as the South Powell

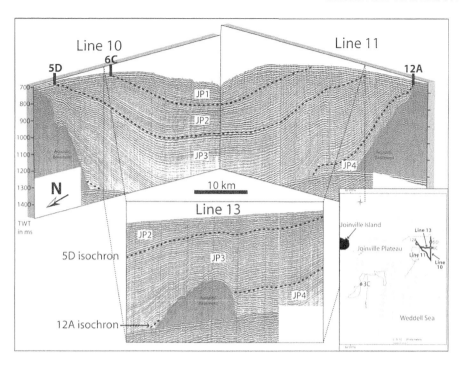

Figure 2. Pseudo three-dimensional image showing correlation of seismic profiles that cross the sedimentary wedge. Seismic units are labeled JP–JP4 (dashed lines are unit boundaries) as they relate to SHALDRIL II cores. The inset map shows the locations of the seismic lines (bold lines) and drill cores (dots).

Ridge [*Maldonado et al.*, 1998] (Figure 1). The ridge is composed of continental blocks, most likely of metamorphic composition, and is considered to be the submarine extension of the Antarctic Peninsula [*Maldonado et al.*, 1998]. The continental slope and rise adjacent to the Joinville Plateau contain a thick sedimentary section and seismic stratigraphic indications of a sediment drift on the continental rise [*Barker and Lonsdale*, 1991].

Seismic stratigraphic investigations of the James Ross Basin (JRB), [*Anderson et al.*, 1992; *Sloan et al.*, 1995; *Smith and Anderson*, 2010] provide an important framework for this investigation. Progradation of the JRB shelf occurred from late Eocene to present. Prominent onlap surfaces within the late Eocene through middle Miocene strata are interpreted to represent eustatic events that are used to derive an age model for the basin [*Smith and Anderson*, 2010]. The seismic records from the southwestern portion of the JRB show thick subglacial and glacimarine units bounded by glacial unconformities. *Smith and Anderson* [2010] conclude that initial expansion of the Antarctic Peninsula Ice Sheet (APIS) onto the eastern continental shelf of the peninsula occurred during the late middle Miocene to early late Miocene and appears to have been contemporaneous with expansion on the western side of the peninsula [*Bart and Anderson*, 1995; *Barker and Camerlenghi*, 2002]. They identified a total of 34 widespread glacial unconformities on the continental shelf, which is similar to the number (31) of glacial unconformities on the western side of the Peninsula [*Bart and Anderson*, 1995].

By late Miocene time, the ice sheet had expanded to the northern part of the peninsula. An expanded section of Plio-Pleistocene strata in the JRB bears a record of increased frequency of ice sheet expansion and retreat cycles (10) during this time, more closely matching the deep-sea oxygen isotope record of ice volume changes than is recorded in the more condensed and tectonically influenced strata on the western side of the peninsula [*Bart and Anderson*, 1995; *Smith and Anderson*, 2010].

The land-based record of glaciation in the northern Antarctic Peninsula has been derived from volcanic and volcaniclastic deposits on James Ross Island and nearby areas [*Smellie et al.*, 2006, 2008; *Hambrey et al.*, 2010]. These results indicate that the ice sheet extended over James Ross and Vega Islands prior to 6.2 Ma and that there were three generally warmer periods at 6.5–5.9, 5.3–4.2, and <0.88 Ma. These same authors argue that the ice sheet was never significantly thicker than it is today, which posses a potential conflict with the marine record of ice sheet grounding on the deep continental shelf [*Smith and Anderson*, 2010].

Strong contour currents generated by the cyclonic motion of the Weddell Gyre dominate the modern oceanographic regime of the northwestern Weddell Sea [*Fahrbach et al.*, 1995; *Von Gyldenfeldt et al.*, 2002]. Currents diverge in the region with a surface current veering to the west across the plateau and into the Bransfield Strait and the other current flowing eastward into Powell Basin (Figure 1). Several investigators have noted the sedimentological influence of these currents in bottom photographs [*Hollister and Elder*, 1969] and in the form of contour current deposits [*Pudsey and Howe*, 1988; *Gilbert et al.*, 1998; *Maldonado et al.*, 2005].

3. DATA AND METHODS

Several single-channel seismic lines were acquired during SHALDRIL II in conjunction with drilling operations as a means of identifying additional targets in more ice-free waters (Figure 1). The data were acquired with a 210 cubic in. air gun source and a shot spacing of 5 s. We recorded a 2 s time interval using a single-channel hydrophone streamer. A band pass filter was applied with a range of 40 to 100 Hz, along with an overall gain function.

SHALDRIL II drill core biostratigraphy [*Bohaty et al.*, this volume] was tied to the seismic data to create isochron horizons used for chronostratigraphic correlation. A minimum velocity function was developed by taking the average interval velocity values of the SHALDRIL II cores recorded by the multisensor core logger and applying a linear relationship between the velocity values with respect to stratigraphic depth. The velocity values were found to be consistent with sonobuoy velocity data from the region [*Hayes*, 1991]. Using the velocity function, the two-way travel time (twt) thickness values were converted to depth.

4. RESULTS

4.1. Stratigraphic Architecture and Seismic Facies

The Joinville Plateau is virtually barren of sediment cover allowing high-resolution imaging of acoustic basement. For the most part, the basement is characterized by a high-amplitude, chaotic reflection pattern with strong diffractions (Figure 3). Coherent reflections occur locally and indicate a possible sedimentary/metasedimentary composition, which is consistent with the dominance of metasedimentary clasts in sediment cores [*Wellner et al.*, this volume].

Southeast of the plateau, there is a thick (greater than 1.5 s twt), southeastward dipping (<4°) wedge of strata that onlaps at high angles onto acoustic basement to the north and west (Figure 2). No definable shelf break (offlap break) is present. The overall seismic character of the stratigraphic succession is highly laminated, with individual reflections showing little thickness variations and with alternating units composed of high and low amplitude reflections. Laminated sediments are correlated across the area with confidence due to the continuous nature of reflections. Condensed sections, marked by onlap surfaces where three or more reflections converge

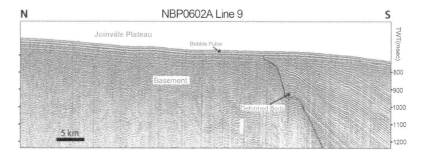

Figure 3. Seismic dip line NBP0602A-9 from SHALDRIL II highlighting the acoustic character of the basement. Note the prominent unconformity that extends across the plateau and sediment wedge. Note also that deformed beds occur only near the basement contact. See Figure 1 for line location.

toward the basement contact, provided drilling targets (Figure 2). These onlap surfaces are not regionally correlative and therefore represent localized thinning, most likely due to tectonic movement of the basement and/or along-strike differences in sedimentation rather than eustasy.

The deepest strata show evidence of some faulting within and just above the basement (Figure 3), most likely associated with basement movement during rifting away of the South Orkney Plateau. Otherwise, the sediment wedge is relatively undeformed. Isolated, internally deformed slumps exhibit tens of meters relief and disrupt the otherwise laminated strata. They occur mainly in the middle part of the section (Figure 4a).

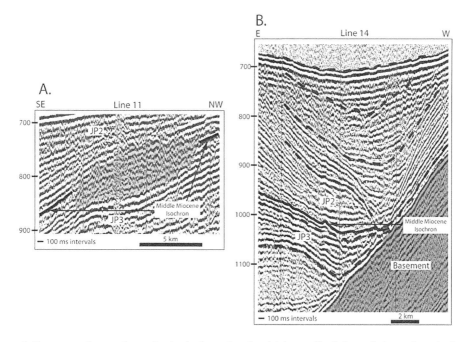

Figure 4. Representative sections of seismic lines showing (a) internally deformed slump deposits in gray shading and (b) canyon fill succession with acoustic basement (including pre-Oligocene strata) shaded for contrast.

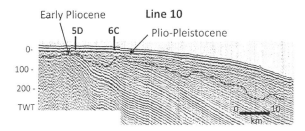

Figure 5. Section of seismic line NBP0602A 10 showing unconformity that truncates strata near the seafloor and extends to approximately 700 m water depth. The age of this surface was constrained by cores from SHALDRIL II sites 5D and 6C.

A prominent canyon exists in the southwestern portion of the study area. The canyon has been largely filled with ~ 400 ms of sediment (Figure 4b). High-angle onlap of canyon walls exists only in the base of the canyon. In the deeper part of the section, the canyon is broad and shallow, approximately 10–15 km wide and 30–50 m deep. In the middle of the section, it narrows and deepens, but then becomes broad and shallow once more near the top of the section. The canyon fill is characterized by low amplitude chaotic/transparent reflections in the narrow and deep portions of the canyon and high amplitude, semicontinuous reflections in the broad and shallow portions of the canyon.

A shallow, relatively narrow channel now occupies the modern seafloor above the canyon (Figure 1). *Gilbert et al.* [1998] first described this feature and interpreted it to be of turbiditic origin, citing evidence from 3.5 kHz seismic profiles and cored sediments from the region (Figure 1). Associated with the channel is a positive bathymetric feature that they interpreted as a sediment drift.

A prominent unconformity extends from the plateau, where acoustic basement is exposed (Figure 3), offshore where it truncates the upper part of the stratigraphic section (Figure 5). It separates a chaotic seismic facies above from the older, acoustically laminated section below and extends to approximately 700 m water depth where it becomes a conformable surface (Figure 5).

4.2. Ages Constraints from Drill Core

4.2.1. Site 12A. The drill site locations were chosen along two seismic lines acquired during SHALDRIL II (Figures 1 and 2). Site 12A was intended to sample the oldest strata in the wedge, above a prominent surface separating strata that are roughly conformable with the basement from laminated sediments that onlap the older strata. The site was selected by mapping the deepest continuous reflections along strike to seismic line 11 where these strata are located near the seafloor (Figure 2). There is no faulting within this part of the stratigraphic section, which gave a preliminary indication that the strata was at least younger than late Eocene/early Oligocene age and postdated rifting between the peninsula and the South Orkney Plateau. Just below the seafloor (0.15–7.2 mbsf), the core sampled sediments containing diatoms and rare calcareous nannofossils that indicate a late Oligocene (~28.4–23.3 Ma) age [*Bohaty et al.*, this volume].

4.2.2. Site 5D. Site 5D cored an interval between 8.0 and 31.4 mbsf, sampling strata above those sampled at Site 12A (Figure 2). On board diatom analysis indicated four diatom zones, representing the Pleistocene, early Pliocene (4.2 and 5.1 M), and middle Miocene (11.7–12.8 Ma) [*Bohaty et al.*, this volume].

Table 1. Average and Maximum Thickness and Sedimentation Rates Computed From Seismic Data and Biostratigraphic Results

Unit	Thickness (twt) (s)	Unit Velocity (m s^{-1})	Thickness Depth (m)	Time Interval (twt)	Sedimentation Rate (cm ka^{-1})
			Average		
JP1	0.1013278	1616.875	81.91719331	4750	1.7
JP2	0.1210053	1776.75	107.4980834	7650	1.4
JP3	0.2940767	1868.25	274.7043974	11600	2.4
			Maximum		
JP1	0.175	1616.875	141.4765625	4750	3.0
JP2	0.35	1776.75	310.93125	7650	4.1
JP3	0.67	1868.25	625.86375	11600	5.4

4.2.3. Site 6C. Site 6C was drilled ~8 km to the south of Site 5D to a depth of 20.5 mbsf. Hole 6C sampled an unconformity that separates late Miocene-early Pliocene sediments below from early Pliocene sediments above [*Bohaty et al.*, this volume].

4.3. Stratigraphic Units and Sedimentation Rates

The Joinville Slope sedimentary wedge can be subdivided into four chronostratigraphic packages based upon the SHALDRIL II drill cores (JP1 through JP4, Figure 2). Each drill core is correlated to the seismic data to create isochrons and used as bounding surfaces for the stratigraphic units. We have assigned the following nomenclature to subdivide the stratigraphic column: JP1, the uppermost unit is bounded by the seafloor at the top and the early Pliocene isochron at the base; JP2, bounded by the early Pliocene isochron at the top and middle Miocene isochron at the base; JP3, bounded by the middle Miocene isochron at the top and late Oligocene isochron at the base; and JP4, bounded by the late Oligocene isochron at the top and the top of basement at the base (Figure 2).

Table 1 shows the regional average and maximum sedimentation rates for each of the seismic stratigraphic units. The average rate is determined by taking the average thickness of each unit. The maximum rate is calculated using the maximum thickness of each unit. The former method yields lower sedimentation rates overall because it includes areas of stratigraphic thinning and condensed intervals. The highest sedimentation rates of ~2.4 cm kyr^{-1} occurred during the late Oligocene to middle Miocene; however, the variation in sedimentation rate is not large (<1 cm kyr^{-1}) and has not varied significantly. Even the maximum sedimentation rates show relatively little variance over time (Table 1).

5. DISCUSSION

Overall, the stratigraphic section of the Joinville Slope is characterized by high-amplitude continuous reflections that correlate across the study area and drape basement highs (Figure 2). This indicates a regional depositional mechanism such as hemipelagic settling and/or sedimentation by contour currents [*Faugères et al.*, 1999]. Additional evidence for contour currents is found in the sedimentological work done on the SHALDRIL II cores. Lithologic descriptions of the cores note an abundance of very fine to fine sand indicating some sorting either before deposition or winnowing of sediments postdeposition [see *Wellner et al.*, this volume]. The SHALDRILL 11 sediment cores also sampled diamictons and pebbly muds [*Wellner et al.*, this volume]. Thus, the section records the combined influence of contour currents and glacimarine processes, which places these strata within the mixed drift system classification of *Rebesco and Stow* [2001].

The role of contour currents in the study area is expected, given the physical oceanographic setting, where the Weddell Gyre and associated branching currents generally follow the curvature of the margin (Figure 1). It is also consistent with results from other studies in the region. To the east of the study area, *Maldonado et al.* [2005] recognized seven styles of contourite drift morphology in the NW Weddell Sea that are mainly controlled by basin physiography. *Pudsey and Howe* [1988] argued that the Weddell Gyre has been the dominant depositional force in the northwestern Weddell Sea region throughout the last few glacial and interglacial cycles. The similar style of stratigraphic architecture and seismic facies indicates that contour currents have influenced sedimentation in the study area since at least the late Oligocene.

Rates of sedimentation on the Joinville slope have not varied significantly, at least not during the Neogene, with average rates ranging from 1.7 to 2.4 cm ka^{-1} and maximum rates ranging from 3.0 to 5.4 cm ka^{-1} (Table 1). These rates are similar to the sedimentation rates reported by *Michels et al.* [2001] for channel levees/drift bodies in the western Weddell Sea (0.7 to 2.8 cm ka^{-1} for interglacial periods and 1.6 to 5.5 cm ka^{-1} for glacial periods).

The main anomaly to the sheet-like deposits of the Joinville slope is the large canyon identified in the southwestern part of the study area. The seismic records provide evidence for periods of incision and fill (Figure 4b). This is shown by the truncation of reflections from previous channel deposits. During periods of deep incision, the canyon was filled with low amplitude, highly chaotic and nearly transparent seismic facies, whereas during less active periods, the canyon was filled with more laminated and high-amplitude seismic facies. Based on the SHALDRIL II sites, episodes of incision and fill occurred during the late Oligocene through middle Miocene, which would have been before the APIS grounded on the continental shelf in the northern peninsula region [*Smith and Anderson*, 2010]. Rather, canyon erosion and turbidite sedimentation were coincident with tectonic movement associated with separation of Joinville Plateau and the South Orkney Plateau.

During the late middle to early late Miocene, the JRB was overridden by an ice sheet with a corresponding increase in rates of margin progradation on the western side of the basin [*Smith and Anderson*, 2010] (Figure 1). There was a corresponding increase in ice rafting recorded in Hole 5D [*Wellner et al.*, this volume] but no evidence for glacial erosion on Joinville Plateau at this time.

A prominent unconformity, probably an amalgamation of several unconformities, extends from the Joinville Plateau across the stratigraphic wedge to approximately 700 m water depth. The unconformity exhibits considerable relief and becomes a conformable surface in a seaward direction (Figure 5). These characteristics are inconsistent with erosion by contour currents. Wind-driven currents are known to influence sedimentation on shallower portions of the Antarctic continental shelf, resulting in gravelly lag deposits at the seafloor, but only to depths of a few hundreds meters [*Anderson*, 1999]. A single swath bathymetry profile across the plateau, where the unconformity is located at or near the seafloor, shows rugged bedrock topography and drumlin-like features (Figure 6). Although the age of these features is not known, they indicate that ice was grounded on the plateau, and the size of these features indicates a prolonged history of glacial sculpturing. These combined observations lead to the interpretation that the unconformity is of glacial origin.

The unconformity was sampled at both Site 5D, where early Pliocene deposits overly middle Miocene deposits, and at Site 6C, where it separates late Miocene-early Pliocene and older strata below from early Pliocene and younger deposits above. Hence, the unconformity was formed

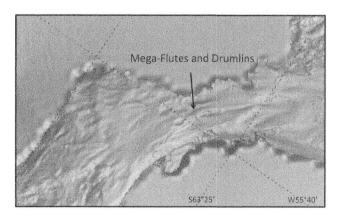

Figure 6. Swath bathymetry image from Joinvile Plateau showing megaflutes and drumlin-like features indicating that ice was grounded on the plateau in the past. Note that this is a single swath so the borders are the edges of the data. See Figure 1 for profile location.

prior to the resumption of early Pliocene sedimentation and is interpreted as recording expansion and grounding of the APIS during the early Pliocene. This interpretation is consistent with the phased northward expansion of the APIS on the eastern side of the Antarctic Peninsula, which began in the late middle Miocene to early late Miocene in the southern part of JRB and expanded onto the northern part of the basin by latest Miocene time [*Smith and Anderson*, 2010]. It is not inconsistent with alpine glaciers having existed in the higher mountains of the peninsula region much earlier [*Birkenmajer*, 1991; *Francis et al.*, 2006; *Ivany et al.*, 2006; *Kirshner and Anderson*, this volume], but does indicate that Joinville Island was likely the last place in Antarctica to be overridden by an ice sheet, and that occurred in the early Pliocene. A number of climate oscillations and associated glacial oscillations [*Smellie et al.*, 2006] and a minimum of 10 expansions of the APIS onto the continental shelf of the James Ross Basin occurred during the late Pliocene-Pleistocene [*Smith and Anderson*, 2010].

6. CONCLUSIONS

The southern flank of the Joinville Plateau has great potential for yielding a high-resolution climate and ice sheet history for the Antarctic Peninsula. Seismic records from the area show a sediment wedge spanning the late Oligocene to the early Pliocene with no apparent hiatuses or large unconformities. The area is also ideal for shallow drilling operations because of the configuration of the sediment wedge geometry oblique to the basement contact. This allows for sampling of the late Oligocene through modern sediments just meters below the seafloor. The regional sedimentation rates calculated using both the drill core and regional seismic lines show little variation in rates of sedimentation throughout the sampled interval with infrequent and isolated episodes of slumping and turbidity current activity. Contour currents associated with the Weddell Gyre have influenced sedimentation in the area since at least the late Oligocene.

Seismic and core evidence suggests that the Antarctic Peninsula Ice Sheet grounded on the Joinville Plateau during the early Pliocene, perhaps following an early Pliocene warm event that is recorded by outcrops on James Ross Island [*Smellie et al.*, 2006].

Acknowledgment. Financial support for this research was provided by the National Science Foundation-Office of Programs (grant 0125922).

REFERENCES

Anderson, J. B. (1999), *Antarctic Marine Geology*, 289 pp., Cambridge Univ. Press, Cambridge, U. K.

Anderson, J. B., S. S. Shipp, and F. P. Siringan (1992), Preliminary seismic stratigraphy of the northwestern Weddell Sea continental shelf, in *Recent Progress in Antarctic Earth Science*, edited by Y. Yoshida et al., pp. 603–612, Terra Sci., Tokyo.

Barker, P. F., and A. Camerlenghi (2002), Synthesis of Leg 178 results: Glacial history of the Antarctic Peninsula from Pacific margin sediments, *Ocean Drill. Program Sci. Results*, *178*, 1–40.

Barker, P. F., and M. J. Lonsdale (1991), A multichannel seismic profile across the Weddell Sea margin of the Antarctic Peninsula: Regional tectonic implications, in *Geological Evolution of Antarctica*, edited by J. A. Thomson et al., pp. 237–241, Cambridge Univ. Press, Cambridge, U. K.

Bart, P. J., and J. B. Anderson (1995), Seismic record of glacial events affecting the Pacific margin of the northwestern Antarctic Peninsula, in *Geology and Seismic Stratigraphy of the Antarctic Margin*, *Antarct. Res. Ser.*, vol. 68, edited by A. K. Cooper, P. F. Barker, and G. Brancolini, pp. 75–96, AGU, Washington, D. C.

Birkenmajer, K. (1991), Tertiary glaciation in the South Shetland Islands, West Antarctica: Evaluation of data, in *Geological Evolution of Antarctica*, edited by M. R. A. Thomson, pp. 627–632, Cambridge Univ. Press, Cambridge, U. K.

Bohaty, S. M., D. K. Kulhanek, S. W. Wise Jr., K. Jemison, S. Warny, and C. Sjunneskog (2011), Age assessment of Eocene–Pliocene drill cores recovered during the SHALDRIL II expedition, Antarctic Peninsula, in *Tectonic, Climatic, and Cryospheric Evolution of the Antarctic Peninsula*, doi:10.1029/2010SP001049, this volume.

Eagles, G., and R. A. Livermore (2002), Opening history of Powell Basin, Antarctic Peninsula, *Mar. Geol.*, *185*, 195–205.

Elliot, D. H. (1988), Tectonic setting and evolution of the James Ross Basin, northern Antarctic Peninsula, in *Geology and Paleontology of Seymour Island, Antarctic Peninsula*, edited by R. M. Feldmann and M. O. Woodburne, *Mem. Geol. Soc. Am.*, *169*, 541–555.

Fahrbach, E., G. Rohardt, N. Scheele, M. Schröder, V. Strass, and A. Wisotzki (1995), Formation and discharge of deep and bottom water in the northwestern Weddell Sea, *J. Mar. Res.*, *53*, 515–538.

Faugères, J.-C., A. V. Dorrik, P. I. Stow, and A. Viana (1999), Seismic features diagnostic of contourite drifts, *Mar. Geol.*, *162*, 1–38.

Francis, J. E., D. Pirrie, and J. A. Crame (Eds.) (2006), *Cretaceous-Tertiary High-Latitude Paleoenvironments, James Ross Basin, Antarctica*, *Geol. Soc. Spec. Publ. 258*, 206 pp.

Gilbert, I. M., C. J. Pudsey, and J. W. Murray (1998), A sediment record of cyclic bottom-current variability from the northwest Weddell Sea, *Sediment. Geol.*, *115*, 185–214.

Hambrey, M. J., J. L. Smellie, A. E. Nelson, and J. S. Johnson (2010), Late Cenozoic glacier-volcano interaction on James Ross Island and adjacent areas, Antarctic Peninsula region, *Geol. Soc. Am. Bull.*, *120*, 709–731.

Hayes, D. E. (Ed.) (1991), *Marine Geological and Geophysical Atlas of the Circum-Antarctic to 30°S*, *Antarct. Res. Ser.*, vol. 54, 56 pp., AGU, Washington, D. C.

Hollister, C. D., and R. B. Elder (1969), Contour currents in the Weddell Sea, *Deep Sea Res. Oceanogr. Abstr.*, *16*, 99–101.

Ivany, L. C., S. van Simaeys, E. W. Domack, and S. D. Samson (2006), Evidence for an early Oligocene ice sheet on the Antarctic Peninsula, *Geology*, *34*, 377–380.

Kirshner, A. E., and J. B. Anderson (2011), Cenozoic glacial history of the northern Antarctic Peninsula: A micromorphological investigation of quartz sand grains, in *Tectonic, Climatic, and Cryospheric Evolution of the Antarctic Peninsula*, doi:10.1029/2010SP001046, this volume.

Lawver, L. A., L. M. Gahagan, and I. W. D. Dalziel (2011), A different look at gateways: Drake Passage and Australia/Antarctica, in *Tectonic, Climatic, and Cryospheric Evolution of the Antarctic Peninsula*, doi:10.1029/2010SP001017, this volume.

Maldonado, A., et al. (1998), Small ocean basin development along the Scotia-Antarctica plate boundary and in the northern Weddell Sea, *Tectonophysics*, *296*, 371–402.

Maldonado, A., et al. (2005), Miocene to recent contourite drifts development in the northern Weddell Sea (Antarctica), *Global Planet. Change*, *45*, 99–129.

Michels, K. H., G. Kuhn, C.-D. Hillenbrand, D. Diekmann, D. K. Fütterer, H. Grobe, and G. Uenzelmann-Neben (2001), The southern Weddell Sea: Combined contourite-turbidite sedimentation at the southeastern margin of the Weddell Gyre, in *Deep-Water Contourite Systems: Modern Drifts and Ancient Series, Seismic and Sedimentary Characteristics*, edited by D. A. V. Stow et al., *Mem. Geol. Soc. London*, *22*, 305–323.

Pudsey, C. J., and J. A. Howe (1988), Quaternary history of the Antarctic Circumpolar Current: Evidence from the Scotia Sea, *Mar. Geol.*, *148*, 83–112.

Rebesco, M., and D. Stow (2001), Seismic expression of contourites and related deposits: A preface, *Mar. Geophys. Res.*, *22*, 303–308.

Sloan, B. J., L. A. Lawver, and J. B. Anderson (1995), Seismic stratigraphy of the Palmer Basin, in *Geology and Seismic Stratigraphy of the Antarctic Margin, Antarct. Res. Ser.*, vol. 68, edited by A. K. Cooper, P. F. Barker, and G. Brancolini, pp. 235–260, AGU, Washington, D. C.

Smellie, J. L., J. M. McArthur, W. C. McIntosh, and R. Esser (2006), Late Neogene interglacial events in the James Ross Island region, *Palaeogeogr. Palaeoclimatol. Palaeoecol.*, *242*, 168–187.

Smellie, J. L., J. S. Johnson, W. C. McIntosh, R. Esser, M. T. Gudmundsson, M. J. Hambrey, and B. van Wyk de Vries (2008), Six million years of glacial history recorded in the James Ross Island Volcanic Group, Antarctic Peninsula, *Palaeogeogr. Palaeoclimatol. Palaeoecol.*, *260*, 122–148.

Smith, R. T., and J. B. Anderson (2010), Ice sheet evolution in James Ross Basin, Weddell Sea margin of the Antarctic Peninsula: The seismic stratigraphic record, *Geol. Soc. Am. Bull.*, *22*, 830–842.

Von Gyldenfeldt, A.-B., E. Fahrbach, M. A. García, and M. Schröder (2002), Flow variability at the tip of the Antarctic Peninsula, *Deep Sea Res., Part II*, *49*, 4743–4766.

Wellner, J. S., J. B. Anderson, W. Ehrmann, F. M. Weaver, A. Kirshner, D. Livsey, and A. Simms (2011), History of an evolving ice sheet as recorded in SHALDRIL cores from the northwestern Weddell Sea, Antarctica, in *Tectonic, Climatic, and Cryospheric Evolution of the Antarctic Peninsula*, doi:10.1029/2010SP001047, this volume.

J. B. Anderson and R. T. Smith, Department of Earth Sciences, Rice University, Houston, TX 77005-1892, USA. (johna@rice.edu)

Age Assessment of Eocene–Pliocene Drill Cores Recovered During the SHALDRIL II Expedition, Antarctic Peninsula

Steven M. Bohaty,[1,2] Denise K. Kulhanek,[3,4] Sherwood W. Wise Jr.,[3] Kelly Jemison,[3] Sophie Warny,[5] and Charlotte Sjunneskog[6]

Pre-Quaternary strata were recovered from four sites on the continental shelf of the eastern Antarctic Peninsula during the SHALDRIL II cruise, NBP0602A (March–April 2006). Fully marine shelf sediments characterize these short cores and contain a mixture of opaline, carbonate-walled, and organic-walled microfossils, suitable for both biostratigraphic and paleoenvironmental studies. Here we compile biostratigraphic information and provide age assessments for the Eocene–Pliocene intervals of these cores, based primarily on diatom biostratigraphy with additional constraints from calcareous nannofossil and dinoflagellate cyst biostratigraphy and strontium isotope dating. The Eocene and Oligocene diatom floras are illustrated in nine figures. A late Eocene age (~37–34 Ma) is assigned to strata recovered in Hole 3C, and a late Oligocene age (~28.4–23.3 Ma) is determined for strata recovered in Hole 12A. Middle Miocene (~12.8–11.7 Ma) and early Pliocene (~5.1–4.3 Ma) ages are assigned to the sequence recovered in holes 5C and 5D, and an early Pliocene age (~5.1–3.8 Ma) is interpreted for cores recovered in holes 6C and 6D. These ages provide chronostratigraphic ground truthing for the thick sequences of Paleogene and Neogene strata present on the northwestern edge of the James Ross Basin and on the northeastern side of the Joinville Plateau, as interpreted from a network of seismic stratigraphic survey lines in the drilling areas. Although representing a coarse-resolution sampling of the complete sedimentary package, the well-constrained ages for these cores also

[1]Earth and Planetary Sciences Department, University of California, Santa Cruz, California, USA.

[2]Now at School of Ocean and Earth Science, University of Southampton, National Oceanography Centre, Southampton, UK.

[3]Department of Earth, Ocean and Atmospheric Sciences, Florida State University, Tallahassee, Florida, USA.

[4]Now at Department of Paleontology, GNS Science, Lower Hutt, New Zealand.

[5]Department of Geology and Geophysics and Museum of Natural Science, Louisiana State University, Baton Rouge, Louisiana, USA.

[6]Antarctic Research Facility, Florida State University, Tallahassee, Florida, USA.

Tectonic, Climatic, and Cryospheric Evolution of the Antarctic Peninsula
Special Publication 063
Copyright 2011 by the American Geophysical Union.
10.1029/2010SP001049

allow for the broad reconstruction of marine and terrestrial paleoenvironments in the Antarctic Peninsula for the late Eocene-to-early Pliocene time
interval.

1. INTRODUCTION

The Antarctic Peninsula is a highly sensitive region to global climate change. Over the past ~50 years, instrumental temperature records from multiple stations show a higher degree of warming in the Antarctic Peninsula relative to other areas of Antarctica and the global average [*Vaughan et al.*, 2003; *Turner et al.*, 2005]. Given the modern sensitivity of this area to atmospheric temper- ature change, a pressing question is the future stability and sensitivity of ice sheets to climate change in the Antarctica Peninsula region. Critical insight into this question can be obtained from reconstructing the past behavior of ice sheets and climatic history of the region. Ice-core records from the Antarctic Peninsula currently extend only back to ~1200 cal yr BP [*Mosley-Thompson and Thompson*, 2003]. Extension of the glacial and paleoclimatic history of the region to millennial and million-year timescales must therefore rely on the geological record.

Most geological information concerning the paleoclimatic history of the Antarctic Peninsula region is confined to late Pleistocene–Holocene time intervals. Marine sediment cores, for example, have been effectively used to reconstruct ice-sheet retreat history since the Last Glacial Maximum [e.g., *Heroy and Anderson*, 2007] and a detailed paleoceanographic history through the Holocene [e.g., *Domack et al.*, 2003]. The pre-Quaternary record is much less known. Knowledge of the Eocene–Pliocene history has been pieced together primarily from drill cores on the continental rise on the western side of the Antarctic Peninsula (Ocean Drilling Program (ODP) Leg 178) [*Barker and Camerlenghi*, 2002] and outcrop records from Brabant, James Ross, Seymour, Cockburn, Alexander, and King George islands [e.g., *Pirrie et al.*, 1997; *Troedson and Smellie*, 2002; *Birkenmajer et al.*, 2005; *Hambrey et al.*, 2008; *Smellie et al.*, 2009]. Obtaining additional records and further exploitation of existing records, however, is necessary to answer key paleoclimatic questions for the region, such as ice sheet presence/stability in warm intervals of the Pliocene [*Smellie et al.*, 2009] and the timing, extent, and character of initial ice-sheet development in the region during the Paleogene [*Anderson et al.*, 2006].

Drilling in high-latitude regions such as the Antarctic Peninsula presents a number of chal- lenges. Most areas along the continental shelves are subject to sea ice cover, drifting icebergs, and rapidly changing weather conditions. In addition, both the coarse-grained nature of proximal glacimarine sediments and the erosive nature of glaciers further hinder the recovery and paleocli- matic interpretation of Antarctic drill cores. In ice-proximal areas, these challenges have been overcome in a variety of ways, including strategic use of large drilling ships by international ocean drilling expeditions (e.g., Deep Sea Drilling Project (DSDP) Leg 28, ODP Legs 178 and 188, and Integrated Ocean Drilling Program Expedition 318) and sea ice/ice shelf-based drilling projects (e.g., Cape Roberts Project and ANDRILL). Nevertheless, drilling in shallow continental shelf areas of the Antarctic Peninsula is not feasible using either a large, non-ice-strengthened drill ship or an ice-based platform. In order to overcome these obstacles, the Shallow Drilling project (SHALDRIL) was designed around the idea of mounting a medium-sized drill rig on an ice- breaking research vessel in order to allow quick drilling of shallow targets in ice-infested areas. Testing of the SHALDRIL strategy was undertaken during two cruises in the austral fall seasons of 2005 and 2006. The primary drilling target of these cruises was a section of thick, seaward dipping strata on the eastern (Weddell Sea) side of the Antarctic Peninsula: an untapped rock sequence ideal for reconstructing the Eocene-to-Pleistocene paleoclimatic and glacial history of the region [*Anderson*, 1999].

The two SHALDRIL cruises utilized the research vessel icebreaker (RVIB) *Nathaniel B. Palmer*, temporarily mounting a drill rig over a moon pool in the aft section of the ship. During the first season of drilling (SHALDRIL Cruise NBP05-02, March–April 2005), a long Holocene succession was cored at a site along the South Shetland Islands [*Milliken et al.*, 2009], as well as a number of technical holes aimed at testing the drill rig (Shipboard Scientific Party, SHALDRIL 2005 cruise report, 2005, http://www.arf.fsu.edu/projects/documents/shaldril2005.pdf). After modification of the drilling tools and techniques, 12 sites were drilled during the second cruise (SHALDRIL II Cruise NBP0602A, March–April 2006) in the northwestern Weddell Sea, off the eastern tip of the Antarctic Peninsula (Shipboard Scientific Party, SHALDRIL II 2006 NBP0602A cruise report, 2006, http://www.arf.fsu.edu/projects/documents/Shaldril_2_Report.pdf, hereinafter Shipboard Scientific Party, 2006). The major successes of this cruise were the recovery of a long Holocene record from the Firth of Tay near Joinville Island [*Michalchuk et al.*, 2009] and penetration into thick, pre-Quaternary strata on the continental shelf at four of the sites: NBP0602A-3, NBP0602A-5, NBP0602A-6, and NBP0602A-12 (Figure 1 and Table 1). Sediments of Eocene age were recovered from a stratigraphic succession within the James Ross Basin at Site 3, and Oligocene–Pliocene age sediments were cored on the northeastern edge of the Joinville Plateau at sites 5, 6, and 12.

In this chapter, we compile and revise the biostratigraphic age constraints for the pre-Quaternary cores recovered on the SHALDRIL II cruise at sites NBP0602A-3, NBP0602A-5, NBP0602A-6, and NBP0602A-12, updating the initial shipboard biostratigraphic report (Shipboard Scientific Party, 2006). We also report on two strontium isotope ages that were generated after the cruise, which provide supporting age information. The presence of reworked sedimentary clasts of Paleocene and Cretaceous age within upper Quaternary sediments recovered at Site NBP0602A-9 has been reported separately by *Kulhanek* [2007]. Although the short cores recovered on the SHALDRIL cruise do not individually span long intervals of time, the age estimates provide valuable constraints for interpreting many different aspects of the geologic and paleoclimate history of the Antarctic Peninsula. Important outcomes of the integrated results include derivation of the erosional and glacial history of the eastern Antarctic Peninsula from the seismic architecture of the James Ross Basin and Joinville Plateau [*Anderson*, 1999; *Smith and*

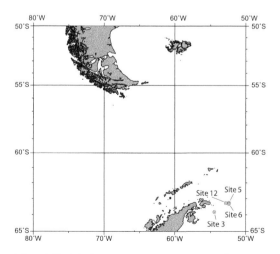

Figure 1. Location map of sites drilled during the SHALDRIL II expedition that recovered pre-Quaternary strata. Drill core sections of Eocene-to-Pliocene age were recovered on the Weddell Sea side of the Antarctic Peninsula at sites NBP0602A-3, NBP0602A-5, NBP0602A-6, and NBP0602A-12.

Table 1. Site Information for Pre-Quaternary Cores Recovered During the SHALDRIL II Cruise

NBP0602A Site	Hole	Location	Latitude/ Longitude	Water Depth[a] (m)	Penetration Depth (m)	Oldest Sediment Recovered
3	C	northern James Ross Basin	63°50.861′S/ 54°39.207′W	340	~20	upper Eocene
5	C	Joinville Plateau	63°15.110′S/ 52°21.908′W	510	11.97	lower Pliocene
5	D	Joinville Plateau	63°15.090′S/ 52°21.939′W	510	31.4	middle Miocene
6	C	Joinville Plateau	63°20.268′S/ 52°22.032′W	531	20.5	upper Miocene(?)– lower Pliocene
6	D	Joinville Plateau	63°19.746′S/ 52°22.040′W	529	~10	lower Pliocene
12	A	Joinville Plateau	63°16.354′S/ 52°49.501′W	442	7.2	upper Oligocene

[a]Drill pipe measurement.

Anderson, 2010, this volume] and the long-term paleovegetation history of the Antarctic Peninsula [*Anderson et al.*, 2011; *Warny and Askin*, this volume(a), this volume(b)].

2. MATERIALS AND METHODS

2.1. Core Material

The drill cores obtained in SHALDRIL II holes 3C, 5C, 5D, 6C, 6D, and 12A are summarized in Figures 2, 3, 4, and 5. A brief overview of the nature of the sediments recovered at each site is given below. A full description and a revised sedimentological summary of the SHALDRIL II cores are presented by *Wellner et al.* [this volume]. Core nomenclature and terminology follows that used in the initial site reports (Shipboard Scientific Party, 2006).

2.1.1. Hole NBP0602A-3C. SHALDRIL II Hole 3C was drilled at a water depth of 340 m on the northern side of the James Ross Basin (Table 1 and Figure 1). Hole 3C was drilled to a depth of ~20 m below seafloor (mbsf), with an average recovery of ~32% (Figure 2). During initial shipboard core description, no lithostratigraphic units were defined for the cores recovered in this hole. The recovered interval (3.0–14.12 mbsf) consists of poorly sorted, muddy very fine to fine sands with a few subangular to subrounded pebbles and shell fragments (bivalve and gastropod) distributed throughout (Shipboard Scientific Party, 2006). The sediments are greenish-black to dark gray in color with no evidence of bioturbation. One large cobble was also noted in core 3C-6R$_a$ (12.35 mbsf), which, along with some of the larger pebbles described throughout the recovered section, contains shell fragments. A shallow shelf depositional environment with minimal glacial influence is interpreted for the strata recovered in this hole [*Wellner et al.*, this volume].

2.1.2. Holes NBP0602A-5C and NBP0602A-5D. SHALDRIL II holes 5C and 5D were drilled at a water depth of 510 m on the northeastern edge of the Joinville Plateau (Table 1 and Figure 1). Overlapping intervals were drilled in the two holes. A highly disturbed, soupy sample was the only core material recovered between 8.5 and 11.97 mbsf in Hole 5C. Hole 5D penetrated to a depth of 31.4 m, with an average recovery of ~40% (Figure 3). Five lithostratigraphic units were defined in Hole 5D (Shipboard Scientific Party, 2006). Subunit IA (8.00–8.80 mbsf) contains a

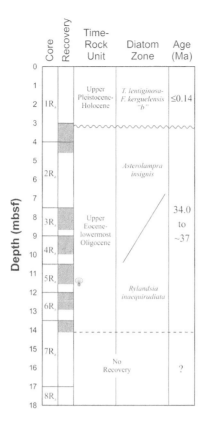

Figure 2. Core recovery (shaded areas), diatom zonal assignments, and age interpretation for Hole NBP0602A-3C. The shell symbol indicates the position of bivalve fragments used for strontium isotope dating.

range of lithologies that include dark greenish-gray sand, dark greenish-gray poorly sorted diamicton, and dark gray muddy sand. Subunit IB (8.80–12.15 mbsf) consists of dark gray muddy sand with no pebbles. Unit II (12.15–15.0 mbsf) consists of olive-green to black sandy diatomaceous mud. Unit III (15.0–22.18) is composed of greenish-gray to black muddy sand with scattered pebbles throughout. Unit IV (22.18–25.07 mbsf) consists of dark gray mud, and unit V (25.07–31.40 mbsf) consists of greenish black sandy mud with pebbles and sandy diatomaceous mud. Units I, II, and the upper part of unit III are interpreted to have been deposited in an outer shelf environment that was influenced by strong contour currents, and the lower part of unit III, unit IV, and unit V are interpreted to have been deposited in a proximal glacimarine shelf environment [*Wellner et al.*, this volume].

2.1.3. Holes NBP0602A-6C and NBP0602A-6D. SHALDRIL II holes 6C and 6D were drilled at ~630 m water depth on the northeastern edge of the Joinville Plateau (Table 1 and Figure 1). Overlapping sections were drilled in the two holes (Figure 4). Hole 6C penetrated to a depth of 20.5 mbsf with an average recovery of 35%, whereas Hole 6D reached a depth of ~10 mbsf with an average recovery of 24%. No lithostratigraphic units were defined for these holes during shipboard work. The sediments are described as greenish-gray pebbly sand, greenish-gray pebbly sandy mud, dark gray to greenish-gray, pebbly muddy sand, and greenish-gray to olive-gray muddy sand (Shipboard Scientific Party, 2006). No macrofossil shell fragments were observed in

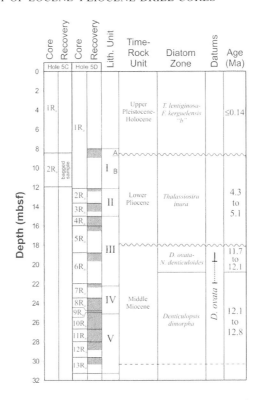

Figure 3. Core recovery (shaded areas), lithostratigraphic units, diatom zonal assignments, and age interpretation for holes NBP0602A-5C and NBP0602A-5D.

the cores from this site. The sandy facies recovered in holes 6C and 6D are interpreted to have been deposited in an outer shelf, current-winnowed environment, similar to the modern-day setting of the Joinville Plateau [*Wellner et al.*, this volume].

2.1.4. Hole NBP0602A-12A. SHALDRIL II Hole 12A was drilled at a water depth of 442 m on the northeastern edge of Joinville Plateau (Table 1 and Figure 1). The hole penetrated to a depth of 7.2 mbsf, with an average recovery of 52% (Figure 5). Six lithostratigraphic units were defined during initial shipboard description of the cores, with lithologies that are predominantly dark gray sandy mud and muddy sand (Shipboard Scientific Party, 2006). Unit I (0.0–0.14 mbsf) consists of black sandy mud with a few pebbles, and unit II (0.14–2.90 mbsf) consists of greenish-gray to black sandy mud with no pebbles. Unit III (2.90 to 3.76 mbsf) is comprised of dark gray muddy sand, and unit IV (3.76–4.20 mbsf) consists of black muddy sand and black sandy mud with small clay lenses throughout. Unit V (4.20–5.25 mbsf) consists of black muddy sand with clay lenses and scattered macrofossil (bivalve) shell fragments, and unit VI (5.35–7.20 mbsf) consists of black sandy mud with rare pebbles and no shell fragments. Units II through VI are interpreted to have been deposited in a distal delta or lower shoreface setting [*Wellner et al.*, this volume].

2.2. Biostratigraphic Ages

Biostratigraphic investigation of cores recovered at SHALDRIL II sites NBP0602A-3, NBP0602A-5, NBP0602A-6, and NBP0602A-12 was focused primarily on diatoms and

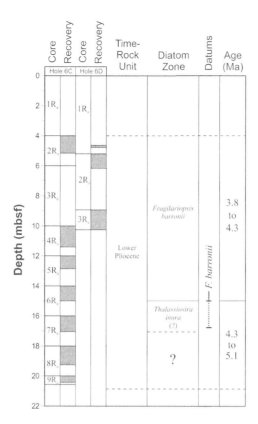

Figure 4. Core recovery (shaded areas), diatom zonal assignments, and age interpretation for holes NBP0602A-6C and NBP0602A-6D.

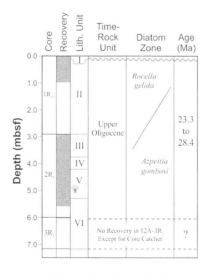

Figure 5. Core recovery (shaded areas), lithostratigraphic units, diatom zonal assignments, and age interpretation for Hole NBP0602A-12A. Note that the upper 13 cm of core 12A-1R$_{a-w}$ is late Pleistocene–Holocene in age. The shell symbol indicates the position of bivalve fragments used for strontium isotope dating.

calcareous nannofossils. Dinoflagellate cysts are also present in these cores and are documented in separate studies by *Warny and Askin* [this volume(a), this volume(b)]. A brief summary of the biostratigraphic age control provided by dinoflagellate cysts for Hole 3C is integrated here with other age information for this section. The biostratigraphic methods for the dinoflagellate cyst work at this site, along with detailed assemblage information for Hole 3C, are included by *Warny and Askin* [this volume(a)].

Qualitative microfossil occurrence data generated in this study are compiled in additional Tables A1–A7, which are electronically archived in the NOAA/World Data Center for Paleoclimatology (ftp://ftp.ncdc.noaa.gov/pub/data/paleo/contributions_by_author/bohaty2011).

Within most intervals of the SHALDRIL II cores, microfossil slide preparations were made from both core-catcher samples and multiple intervals within each section. Since the sequences drilled at these sites are relatively short cores from thick Paleogene and Neogene sequences, few biostratigraphic events (first occurrence or last occurrence datum levels) were identified within continuous intervals of the cores. Therefore, biostratigraphic age constraints rely on the assemblage characteristics and the presence/absence of selected taxa; biostratigraphic interpretations based on the absence of particular taxa are made only for taxa of known paleoecology that are typically common at high southern latitude sites. For Neogene diatom datums, we preferentially use age calibrations from sites located at similar latitudes and environments to SHALDRIL II cores (e.g., sites drilled during ODP Legs 113 (Maud Rise), 188 (Prydz Bay), and 178 (Antarctic Peninsula)), rather than the Southern Ocean compilation and statistical age calibrations of *Cody et al.* [2008]. Ages for all datum events are calibrated to the *Gradstein et al.* [2004] timescale. In most cases, determination of the revised age calibrations required updating published ages originally calibrated using the geomagnetic polarity timescale of *Cande and Kent* [1992, 1995] and Cenozoic global chronostratigraphic compilation of *Berggren et al.* [1995].

2.3. Diatom Methods and Zonal Schemes

Samples from the SHALDRIL II cores were prepared for diatom analysis using standard procedures. A smear or strewn slide was initially prepared for all samples and examined under a light microscope. If necessary, the samples were further prepared using chemical treatment and/or sieving. The chemically treated samples were reacted in small beakers with 10% hydrochloric acid in order to remove the carbonate component, followed by 10% hydrogen peroxide to remove labile organic matter. The samples were not heated during chemical treatment. Selected samples were also sieved at 10 μm using nylon screens to concentrate larger taxa. All samples were prepared using 20 × 40 mm coverslips and mounted using Norland Optical Adhesive #61 (refractive index of 1.56). The slides were examined on a Zeiss Axioscope microscope at 400× and 1000×, with the higher power used mainly for taxonomic identification.

Relative diatom abundance was determined qualitatively from smear or strewn slides of unsieved preparations. The total relative abundance of diatoms (as a group) was determined at 400× magnification and was based on the average number of specimens observed per field of view. Several traverses were made across the coverslip, and abundance estimates were recorded as follows: A (abundant), >10 valves per field of view; C (common), 3–9 valves per field of view; F (few), 1 to 2 valve(s) per field of view; R (rare), 1 valve in 2–30 fields of view; T (trace), very rare fragments present; and B (barren), no diatom valves or fragments present.

The qualitative abundance of individual diatom taxa was based on the number of specimens observed per field of view at 1000× (oil objective). Individual species abundance categories are listed below. These abundance categories are comparable to other studies that have qualitatively documented diatom occurrence in Antarctic shelf sediments [e.g., *Harwood*, 1989; *Scherer et al.*, 2000; *Winter and Iwai*, 2002]. Generally, one quarter to one half of the 20 × 40 mm coverslip was examined (40 mm = ~200 fields of view). After initial abundance determinations were made at 1000×, the slides were then routinely scanned at 400× to identify rare taxa. Abbreviations are as follows: A (abundant), ≥2 valves per field of view; C (common), 1 to 5 valve(s) in 5 fields of view; F (few), ~1 to 3 valve(s) in 20 fields of view; R (rare), ~1 to 2 valve(s) in 60 fields of view; X (present), ≤1 valve or identifiable fragment per traverse of coverslip; *r*, rare occurrences of a taxon interpreted as reworked specimens; *d*, rare occurrences of a taxon interpreted as downcore contamination; and "?," uncertain identification of a particular taxon.

The degree of siliceous microfossil fragmentation often mirrors dissolution, but the two factors are not necessarily dependent (i.e., well-preserved samples can be highly fragmented). Preservation of diatoms, therefore, was qualitatively based on the degree of dissolution and was rated as follows: G (good), slight to no dissolution; M (moderate), moderate dissolution; and P (poor), severe effects of dissolution. In addition, the degree of fragmentation was also noted: L (low), minimal fragmentation; M (moderate), valves moderately fragmented; and H (high), highly fragmented with very few complete valves present.

The primary goal of diatom analysis in this study was to identify the presence/absence of important marker taxa. Of particular relevance to the biostratigraphic dating of the SHALDRIL II cores are the diatom studies of cores obtained from the western side of the Antarctic Peninsula during ODP Leg 178 [*Iwai and Winter*, 2002; *Winter and Iwai*, 2002]. Outside of the Antarctic Peninsula region, a number of extensive diatom biostratigraphic studies have been also carried out for Cenozoic cores recovered at pelagic sites around the Southern Ocean [e.g., *Schrader*, 1976; *Gersonde and Burckle*, 1990; *Baldauf and Barron*, 1991; *Harwood and Maruyama*, 1992; *Gersonde and Bárcena*, 1998; *Censarek and Gersonde*, 2002; *Zielinski and Gersonde*, 2002]. These studies have resulted in the age calibration of numerous diatom taxa and several proposed zonal schemes. Currently, detailed and well-calibrated Southern Ocean zonal schemes only exist for the Oligocene-to-Pleistocene time interval. The schemes utilized here for the SHALDRIL II cores are drawn primarily from three sources: *Harwood and Maruyama* [1992], *Censarek and Gersonde* [2002], and *Zielinski and Gersonde* [2002].

The application of the standard Southern Ocean diatom zonal schemes to Antarctic shelf sections is problematic. Antarctic shelf diatom assemblages, such as those observed in the SHALDRIL II cores, are typically very different than open-ocean assemblages, and many of the marker taxa that are biostratigraphically useful at deep-sea locations are either not present or are present in low abundances in coastal/neritic shelf assemblages. Although the presence of pelagic taxa is sufficient to apply the deep sea-derived zonal schemes to the SHALDRIL II cores, refined age calibrations and zonations for application in shelf areas are currently in development [e.g., *Olney et al.*, 2007; *Winter et al.*, 2011a].

2.4. Calcareous Nannofossil Methods

Smear slides were prepared for calcareous nannofossil study from SHALDRIL II cores using standard techniques. Several strewn slides were also made from Hole 3C samples due to the sandy lithologies recovered in this section. The nannofossil slides were examined using a Zeiss Axioscope microscope under cross-polarized, plain-transmitted, and phase-contrast light at 1000×–1200× magnification. Species preservation and abundance vary significantly due to etching, dissolution, or calcite overgrowth. Six calcareous nannofossil abundance levels were designated as follows: A (abundant), 1–10 specimens per field of view; C (common), 1 specimen per 2–10 fields of view; F (few), 1 specimen per 11–100 fields of view; R (rare), 1 specimen per >100 fields of view; tr (trace), possible calcareous nannofossil(s) but unidentifiable to genus or species level; and B (barren), no nannofossils observed in the sample.

Nannofossil preservation was evaluated as follows: G (good), little or no evidence of dissolution and/or overgrowth, primary morphological characteristics only slightly altered, specimens identifiable to the species level; M (moderate), specimens exhibiting some etching and/or overgrowth, primary morphological characteristics sometimes altered but most specimens identifiable to the species level; and P (poor), specimens severely etched or exhibiting overgrowth, primary morphological characteristics largely destroyed, specimens cannot be identified at the species and/or generic level.

2.5. Strontium Isotope Methods

In the course of core description on board the RVIB *Nathaniel B. Palmer*, several bivalve fragments were identified and picked from cores 3C-5R$_a$ and 12A-2R$_a$ for strontium isotope dating. The samples were crushed to fragments, lightly leached with acetic acid, washed in ultrapure water, and dried in a clean environment. Clean fragments were handpicked under the microscope, and analysis using X-ray diffraction (XRD) showed the samples to be 100% aragonite. Strontium separation was carried out using routine ion-exchange methods, and strontium isotope analysis was performed by standard protocols using thermal ionization mass spectrometry (see *McArthur et al.* [2006] for further details).

Strontium isotope ratios were determined for only two samples from holes 3C and 12A. Although a robust strontium-isotope age assessment of these cores would require analysis of numerous specimens and, ideally, a range of sample depths, we include these results here as support for the biostratigraphic age interpretations. The strontium isotope age interpretations and uncertainty estimates for these samples were derived using the LOWESS fit to the Eocene–Oligocene compilation of multiple strontium isotope records, adjusted to the *Gradstein et al.* [2004] timescale [*McArthur and Howarth*, 2004].

3. RESULTS

Diatoms are present in most intervals of SHALDRIL holes 3C, 5C, 5D, 6C, 6D, and 12A. Qualitative diatom census data for these drill cores are presented in additional Tables A1, A2, A4, A5, A6, and A7. In most sections, good diatom preservation and the presence of pelagic marker taxa allow assignment of relatively well constrained biostratigraphic ages. Interpreted diatom biostratigraphic ages and assigned zones are summarized in Figures 2, 3, 4, 5, 6, and 7.

Calcareous nannofossils are present and well preserved in most samples from Hole 3C, sporadically present in Hole 12A, and absent in Neogene strata recovered in holes 5C, 5D, 6C, and 6D. A checklist and qualitative abundance assessment of nannofossil taxa observed in holes 3C and 12A are presented in Table A3. Although few zonal marker taxa were identified in holes 3C and 12A, calcareous nannofossil biostratigraphy does allow interpretation of broad age constraints.

Dinoflagellate cysts are present and well preserved in all of the pre-Quaternary sections recovered by the SHALDRIL II program [*Warny and Askin*, this volume(a), this volume(b)], but only Hole 3C contains age-diagnostic dinoflagellate cyst taxa [*Warny and Askin*, this volume(a)]. A brief summary of the age control provided by these taxa is included below within the biostratigraphic assessment of Hole 3C.

Strontium isotope analysis of the bivalve fragments from cores 3C-5R$_a$ and 12A-2R$_a$ yielded $^{87}Sr/^{86}Sr$ ratios (Table 2) that fall within the range of seawater values recorded in the marine strontium isotope curve for the middle Eocene-to-late Oligocene time interval (0.70770–0.70820)

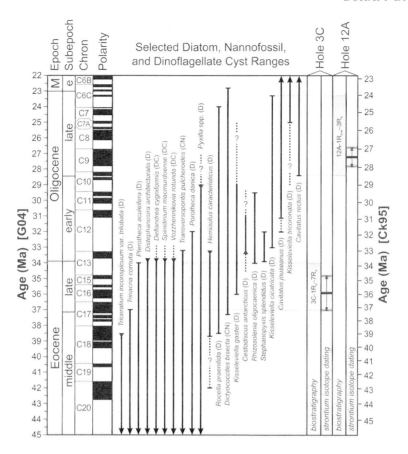

Figure 6. Summary of age interpretations for Eocene–Oligocene cores recovered at sites 3 and 12. The ranges of key taxa providing biostratigraphic age constraint are plotted, and the fossil group of each taxon is labeled as follows: D, diatom; CN, calcareous nannofossil; DC, dinoflagellate cyst. The shaded fields in the biostratigraphic age columns represent the narrowest/shortest intervals for each section obtained by applying diatom, dinoflagellate cyst, and nannofossil age constraints. In the adjacent columns, the strontium isotope ages are indicated with error estimates derived from the LOWESS calibration fit of the marine $^{87}Sr/^{86}Sr$ curve [*McArthur and Howarth*, 2004].

[*McArthur and Howarth*, 2004]. These results, in conjunction with the pristine visual appearance and aragonitic mineralogy of the analyzed specimens, suggest that these ratios record a primary seawater signal. We therefore use these data to calculate ages via comparison with the Paleogene strontium isotope curve [*McArthur and Howarth*, 2004] and compare these strontium isotope results with the biostratigraphic age interpretations.

4. BIOSTRATIGRAPHIC AND STRONTIUM ISOTOPE AGE ASSESSMENTS

4.1. Hole NBP0602A-3C

4.1.1. Diatoms. Diatoms from the uppermost part of core 3C-1R$_a$ (samples 3C-1R$_a$-1, 0 cm and 3C-1R$_a$-1, 15 cm; 3.00 to 3.15 mbsf) are well preserved and represent a modern sea ice assemblage (Table A1). The samples are dominated by *Chaetoceros* spp. and *Fragilariopsis curta* with secondary abundances of *Fragilariopsis cylindrus* and *Thalassiosira antarctica*. This

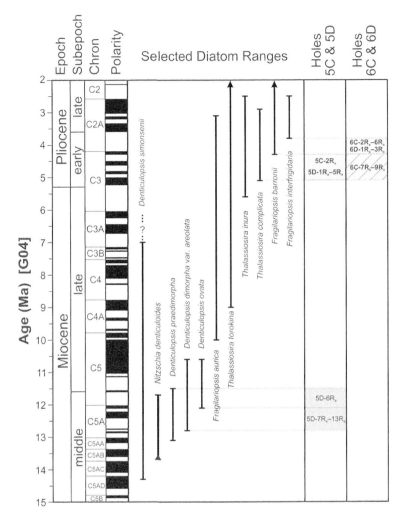

Figure 7. Summary of age interpretations for Miocene–Pliocene cores recovered at sites 5 and 6. The shaded fields in the biostratigraphic age columns represent the narrowest/shortest intervals for each section obtained by applying diatom age constraints. The ages of several cores at the bottom of Hole 5C are not well constrained due to poor diatom preservation; the broad age constraints for these cores are indicated with shaded areas filled by diagonal lines.

uppermost interval of the drilled section is presumed to be Holocene in age, but diatom biostratigraphy only constrains the age of the samples to ~140 ka or younger based on the absence of *Rouxia leventerae* [*Zielinski et al.*, 2002].

A distinct downcore change in diatom assemblages occurs in the lower part of core 3C-1R_a. Sample 3C-1R_a-CC (4.0 mbsf) contains very rare and poorly preserved diatoms, but the presence of *Pseudorutilaria* spp., *Pterotheca* spp., *Pyxilla* spp., and large, heavily silicified *Stephanopyxis* spp. (Table A2) clearly indicates that Paleogene strata were penetrated at very shallow depths in the hole. Below this level, a relatively uniform diatom assemblage is present down to the base of the recovered interval in core 3C-7R_a (~14.1 mbsf). Diatoms are rare and poorly preserved in the upper part of this interval (cores 1R_a to 3R_a; ~4.0 to 9.0 mbsf), whereas diatoms in the lower

Table 2. Stronitum Isotope Results From SHALDRIL II Cores

Hole	Core	Section/ Interval	Depth (mbsf)	Material	$^{87}Sr/^{86}Sr$	LOWESS Age[a] (Ma)	Age Uncertainty (± Myr)
3C	5R$_a$	CC	11.56	bivalve fragment, aragonite	0.707763	35.9	1.1
12A	2R$_a$	2/206 cm	4.96	bivalve fragment, aragonite	0.708085	27.2	0.6

[a]*McArthur and Howarth* [2004].

interval (cores 3C-4R$_a$ to 3C-7R$_a$; ~9.0 to 14.1 mbsf) are few to common in abundance and moderately well preserved. Diatom assemblages from cores 3C-6R$_a$ and 3C-7R$_a$ are particularly diverse and well preserved. Overall, the diatom assemblages throughout the Paleogene interval of Hole 3C are dominated by neritic-planktic taxa such as *Radialiplicata clavigera*, *Stephanopyxis* spp., *Stictodiscus* spp., *Pseudorutilaria* spp., and *Pterotheca* spp. (Table A2). These groups are characteristic of Eocene–Oligocene neritic shelf facies from the Antarctic margin [e.g., *Harwood*, 1989; *Harwood and Bohaty*, 2001].

Although the Paleogene diatom assemblages from cores 3C-1R$_a$ to 3C-7R$_a$ are dominated by numerous shelfal taxa that are not biostratigraphically useful, the presence of several less abundant taxa provide a rough age constraint for the section. These taxa include *Cestodiscus convexus*, *Eurossia irregularis*, *Goniothecium rogersii*, *Hemiaulus dissimilis*, *Hemiaulus reflexispinosus*, *Porotheca danica*, *Pyxilla* spp., *Rocella praenitida*, and *Trochosira spinosus* (Table A2). These taxa are common in Paleogene sections around the Antarctic margin, and their documented stratigraphic ranges [*Hajós*, 1976; *Schrader*, 1976; *Gombos*, 1983; *Gombos and Ciesielski*, 1983; *Fenner*, 1984; *Harwood*, 1989; *Scherer et al.*, 2000; *Harwood and Bohaty*, 2001; *Strelnikova et al.*, 2001; *Olney et al.*, 2007] provide a general age assignment of late middle Eocene to early Oligocene for the Hole 3C drill core. The presence of *Pyxilla* spp. and *R. praenitida*, in particular, supports this broad age assignment. *Pyxilla* spp. typically only occur as rare or reworked specimens in late Oligocene age or younger sediments in the Antarctic region and therefore indicate an age of early Oligocene or older (≥~28 Ma, Table 3) for the Hole 3C section. A maximum age of ~38.1 Ma can be inferred based on the first occurrence (FO) of *R. praenitida* at ODP Site 1172 (C. Stickley, personal communication, 2010) (Table 3). This maximum age estimate is further supported by the absence of *Triceratium inconspicuum* var. *trilobata*, a cosmopolitan species characteristic of middle Eocene diatom assemblages [*Gombos*, 1983; *Fenner*, 1984, 1985], which also suggests a late middle Eocene or younger age (≤38.5 Ma, Table 3) for Hole 3C.

Although the diatom assemblages from Hole 3C are relatively diverse, a precise biostratigraphic age estimate is difficult due to the fact that very few sections of late Eocene–early Oligocene age have been recovered from the Antarctic shelf, and thus, the age ranges for most diatom taxa in this interval are not well calibrated. The presence or absence of several additional taxa, however, allows refinement of the broad age estimate provided by diatoms for this section (Table 3). An Eocene age is strongly supported by the presence of several taxa, including *Distephanosira architecturalis*, *Hemiaulus caracteristicus*, and *Pterotheca aculeifera*. All three of these taxa have last occurrence (LO) or last common occurrence (LCO) datums near the Eocene–Oligocene boundary, thus indicating an age older than 34.0 Ma (Table 3). Additionally, several characteristic early Oligocene diatom taxa are also missing in the Hole 3C assemblage, including *Cestodiscus antarcticus*, *Rhizosolenia oligocaenica*, *Skeletonemopsis mahoodii*, and *Stephanopyxis splendidus*. All of these taxa have FO or first common occurrence (FCO) datums

Table 3. Bioevents Used to Constrain the Ages of SHALDRIL II Holes 3C and 12A

Hole	Fossil Group	Present/Absent	Event	Taxon	Uppermost Level Present (mbsf)	Lowermost Level Present (mbsf)	Age[a] (Ma)	Age[b] (Ma)	Chron	Age Calibration Source(s)[c]	Age Calibration Site(s)	Interpreted Age Constraint (Ma)
3C	diatom	present	LCO	*Pyxilla* spp.	4.00	14.12	~27–29	~27–29	C9n–C11n.2n	GC, BB, HM	513, 744, 748	≥~27–29
3C	diatom	absent	FO	*C. jouseanus* (early form)	-	-	~32	~32	C12r	HB	CRP-3	≥~32
3C	diatom	present	LO	*P. danica*	12.92	14.42	~32	~32	C12r	GC, S	511[d] 513, 1172	≥~32
3C	nannofossil	present	LO	*T. pulcheroides*	12.87	14.12	33.0	33.2	top C13n	Ws	511[d]	≥33.2
3C	diatom	present	LCO	*H. caracteristicus*	9.97	14.42	33.0	33.2	top C13n	GC	511[d]	≥33.2
3C	diatom	absent	FCO	*Cestodiscus antarcticus*	-	-	33.2	33.4	C13n	H+	689	≥33.4
3C	dinoflagellate	present	LCO	*Deflandrea cygniformis*	4.30	14.10	33.5	33.7	C13n/C13r	WB, SH	511, 1172	≥33.7
3C	dinoflagellate	present	LCO	*Spinidinium macmurdoense*	3.54	14.10	33.5	33.7	C13n/C13r	WB, SH	511, 1172	≥33.7
3C	dinoflagellate	present	LCO	*Vozzhennikovia rotunda*	3.54	13.60	33.5	33.7	C13n/C13r	SH	511, 1172	≥33.7
3C	diatom	present	LCO	*D. architecturalis*	8.68	14.12	33.5	33.7	C13n/C13r	GC, H+	511,[d] 689	≥33.7
3C	diatom	absent	FO	*Stephanopyxis splendidus*	-	-	33.7	33.9	C13r	GC	511[d]	≥33.9
3C	diatom	present	LO	*P. aculeifera*	4.57	14.12	33.8	34.0	C13r	GC	511[d]	≥34.0

3C	diatom	FO	*K. gaster*	present	8.68	~36	~36	13.60	C16n.2n	Hj, Hr	280, CIROS-1	≤~36
3C	diatom	LO	*Trinacria cornuta*	absent	-	~37	~37	-	C17n	S	1172	≤~37
3C	diatom	FO	*R. praeniitida*	present	8.68	~38.6	~38.1	14.12	C18n	S	1172	≤~38.1
3C	diatom	LCO	*Triceratium inconspicuum* var. *trilobata*	absent	-	~39	~38.5	-	C18n	S	1172	≤~38.5
3C	diatom	FO	*H. caracteristicus*	present	9.97	~43	~42	14.42	C20n	S	1172	≤~42
12A	diatom	LO	*K. tricoronata*	present	0.30	~21.5	~20.9	4.36	C6Ar	Sc, Wi, O	CRP-2/2A	≥~20.9
12A	nannofossil	LO	*D. bisectus*	present	0.96	23.6	22.8	0.96	C6Cn.1r	P, Br, W	522, 703, 558, 563	≥22.8
12A	diatom	LO	*K. cicatricata*	present	0.30	24.0	23.3	4.36	C6Cn.3n	Sc, W, O	CRP-2/2A	≥23.3
12A	diatom	FO	*C. rectus*	present	0.30	28.5	28.4	3.60	C10n.1n	B	1220	≤28.4
12A	diatom	FO	*C. jouseanus* (sensu stricto)	present	0.30	30.9	31.1	4.36	C12n	R	744, 748	≤31.1

[a] *Cande and Kent* [1995] and *Berggren et al.* [1995].

[b] *Gradstein et al.* [2004].

[c] Abbreviations are as follows: Hj, *Hajós* [1976]; P, *Poore et al.* [1983]; Ws, *Wise* [1983]; GC, *Gombos and Ciesielski* [1983]; Hr, *Harwood* [1989]; BB, *Baldauf and Barron* [1991]; HM, *Harwood and Maruyama* [1992]; Br, *Berggren et al.* [1995]; Sc, *Scherer et al.* [2000]; Wi, *Wilson et al.* [2000]; HB, *Harwood and Bohaty* [2001]; W, *Wilson et al.* [2002]; B, *Barron et al.* [2004]; R, *Roberts et al.* [2003]; WB, *Williams et al.* [2004]; O, *Olney et al.* [2007]; H+, D. Harwood and S. Bohaty (unpublished data, 2010); S. C. Stickley (unpublished data, 2010); SH, S. Houben (unpublished data, 2010).

[d] Using revised age model for Hole 511 (S. Bohaty and S. Houben, unpublished data, 2010).

near the Eocene–Oligocene boundary [*Gombos and Ciesielski*, 1983; *Fenner*, 1984; *Roberts et al.*, 2003]. If the absence of these taxa in Hole 3C is considered to be biostratigraphically significant, then an Eocene age is also supported. One caveat is that *C. antarcticus* and *R. oligocaenica* are typically rare in shelf sediments, and their absence may be due to environmental factors. *S. mahoodii* and *S. splendidus*, on the other hand, are relatively common in early Oligocene shelf sediments [*Barron and Mahood*, 1993; *Sims*, 1994; *Scherer et al.*, 2000; *Harwood and Bohaty*, 2001]; this suggests their absence in Hole 3C is not environmentally controlled, thus supporting an Eocene age for the Hole 3C section.

A more precise maximum age estimate for Hole 3C beyond that given by the presence of *R. praenitida* (≤38.1 Ma) and absence of *T. inconspicuum* var. *trilobata* (≤38.5 Ma) is difficult, due to the lack of continuous and well-studied middle-to-late Eocene sections in the southern high latitudes. The presence of *Kisseleviella gaster* can be used to infer a late Eocene or younger age, although the FO datum of this taxon is only approximately constrained. *K. gaster* is documented in upper Eocene–lower Oligocene sediments of DSDP Hole 280 [*Hajós*, 1976] and in upper Eocene sediments of the CIROS-1 drill core [*Harwood*, 1989] and ODP Hole 1166A (S. Bohaty, unpublished data), and *K. gaster* is absent in middle Eocene shelf facies of ODP Hole 1172A that were deposited prior to major deepening of the site in the late Eocene (C. Stickley, personal communication, 2010). Therefore, a conservative interpretation of the biostratigraphic constraint provided by the FO of *K. gaster* is that the Hole 3C section is no older than ~37 Ma. Additional support for this interpretation is provided by the absence of *Trinacria cornuta* in Hole 3C. The LCO of *T. cornuta* is also not well constrained, but is placed tentatively at ~37 Ma (Subchron C17n) in Hole 1172A (C. Stickley, personal communication, 2010).

In summary, diatom biostratigraphy indicates that the Paleogene section recovered in Hole 3C was deposited between ~37 and 34 Ma (late Eocene) based on the presence/absence of several taxa (Table 3). This interval corresponds to the *Rylandsia inaequiradiata* and *Asterolampra insignis* zones of *Gombos and Ciesielski* [1983] (Figure 2), although the zonal marker for the base of this interval (*R. inaequiradiata*) was not identified in Hole 3C samples. Further refinement of the age of this section with diatom biostratigraphy will most likely be possible in the future as more Eocene–Oligocene age sections are recovered from the Antarctic region. A late Eocene diatom assemblage similar to that present in Hole 3C has previously only been documented in a few Southern Ocean drill cores [e.g., *Hajós*, 1976; *Gombos and Ciesielski*, 1983], although the benthic and neritic components of the assemblages in Hole 3C share an affinity with late Eocene diatom assemblages from the Oamaru diatomite in New Zealand [*Desikachary and Sreelatha*, 1989; *Edwards*, 1991]. A selection of diatom taxa present in the Hole 3C assemblage are illustrated in Figures 8–15.

4.1.2. Calcareous Nannofossils. Calcareous nannofossils are present in nearly all samples examined from Hole 3C (cores 3C-1R$_a$ to 3C-7R$_a$, Table A3). Most specimens are well preserved and are rare to locally abundant from a paleontologic point of view, although present in only trace amounts as a lithologic component.

Most Hole 3C samples are dominated by small-to-medium size reticulofenestrids, including *Reticulofenestra minuta*, *Reticulofenestra minutula*, and *Reticulofenestra daviesii* (Table A3). *Braarudosphaera bigelowii* is also common throughout the section, typically in low numbers, although abundant in sample 3C-7R$_a$-1, 30 cm (13.80 mbsf). This species is a salinity-tolerant encystment form common in continental margin environments [*Bukry*, 1974; *Perch-Nielsen*, 1985]. Many pristine specimens of both *B. bigelowii* and *R. minutula* were observed in several samples, often as complete coccospheres, indicating diminished bioturbation rates in the organic-rich sediments deposited at this site. Reworked Cretaceous species are found

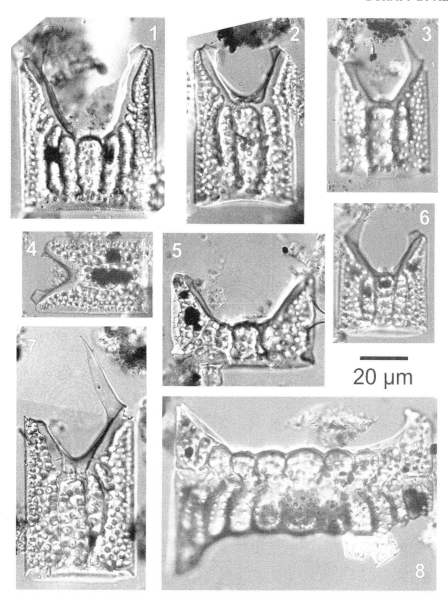

Figure 8. Light microscope images of upper Eocene diatoms from Hole 3C. Scale bar equals 20 μm. Numbers 1–8 show *Hemiaulus reflexispinosus* Ross and Sims, samples: 1, 3C-6R$_a$, CC; 2, 3C-6R$_a$, CC; 3, 3C-6R$_a$, CC; 4, 3C-6R$_a$, CC; 5, 3C-6R$_a$, CC; 6, 3C-7R$_a$-1, 10 cm; 7, 3C-6R$_a$, CC; 8, 3C-7R$_a$-1, 10 cm.

sporadically in the upper 8 m of the hole (cores 3C-1R$_a$ to 3C-3R$_a$, Table A3). Additionally, *Markalius apertus* and *Toweius pertusus*, characteristic of the Paleocene and Paleocene–early Eocene, respectively, occur sporadically throughout the section, indicative of the additional presence of early Paleogene reworked material.

Owing to the low diversity of calcareous nannofossils identified in Hole 3C, it is difficult to independently establish a precise biostratigraphic age for the section using nannofossils. The presence of *R. daviesii*, *Zygrhablithus bijugatus*, and *Cyclicargolithus floridanus* together are

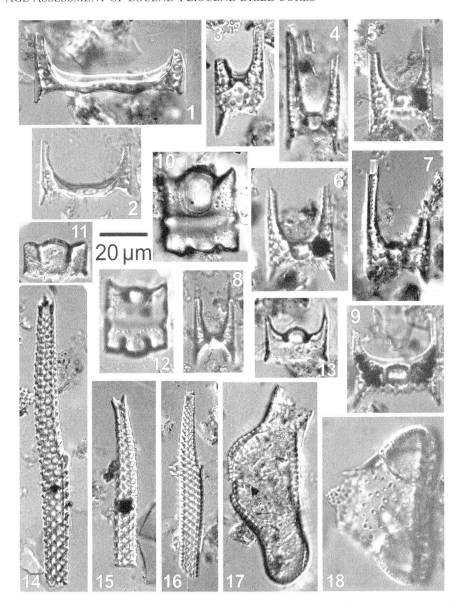

Figure 9. Light microscope images of upper Eocene diatoms from Hole 3C. Scale bar equals 20 μm. Number 1 shows *Hemiaulus* cf. *caracteristicus* Hajós, sample 3C-6R$_a$, CC. Number 2 shows *H. caracteristicus* Hajós, sample 3C-6R$_a$, CC. Numbers 3–9 show *Hemiaulus* sp. 1 (this study), samples: 3, 3C-6R$_a$, CC; 4, 3C-6R$_a$, CC; 5, 3C-6R$_a$, CC; 6, 3C-7R$_a$, CC; 7, 3C-6R$_a$, CC; 8, 3C-7R$_a$, CC; 9, 3C-6R$_a$, CC. Numbers 10–13 show *Hemiaulus dissimilis* Grove and Sturt, samples: 10, 3C-6R$_a$, CC; 11, 3C-6R$_a$, CC; 12, 3C-6R$_a$, CC; 13, 3C-6R$_a$, CC. Numbers 14–16 show *Pyxilla reticulata* Grove and Sturt, samples: 14, 3C-6R$_a$, CC; 15, 3C-6R$_a$, CC; 16, 3C-6R$_a$, CC. Number 17 shows *Biddulphia* (?) sp., sample 3C-6R$_a$, CC. Number 18 shows "*Triceratium*" sp., sample 3C-6R$_a$, CC.

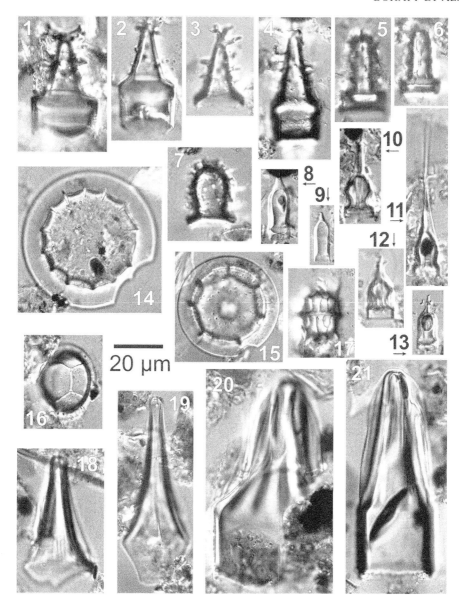

Figure 10. Light microscope images of upper Eocene diatoms from Hole 3C. Scale bar equals 20 μm. Numbers 1–4 show *Syndendrium rugosum* Suto, samples: 1, 3C-6R$_a$, CC; 2, 3C-6R$_a$, CC; 3, 3C-6R$_a$, CC; 4, 3C-6R$_a$, CC. Numbers 5 and 6 show *Pterotheca* sp. 2 (this study), samples: 5, 3C-6R$_a$, CC; 6, 3C-6R$_a$, CC. Number 7 shows *Syndendrium* sp., sample 3C-6R$_a$, CC. Numbers 8 and 9 show *Pseudopyxilla* sp., samples: 8, 3C-6R$_a$, CC; 9, 3C-6R$_a$, CC. Numbers 10–13 show *Pterotheca aculeifera* Grunow, samples: 10, 3C-6R$_a$, CC; 11, 3C-6R$_a$, CC; 12, 3C-6R$_a$, CC; 13, 3C-6R$_a$, CC. Numbers 14–16 show "*Poretzkia*? sp." of *Hajós* [1976], samples: 14, 3C-6R$_a$, CC; 15, 3C-6R$_a$, CC; 16, oblique view, 3C-6R$_a$, CC. Number 17 shows *Stephanogonia* sp., sample 3C-6R$_a$, CC. Numbers 18–21 show *Porotheca danica* (Grunow) Fenner, samples: 18, 3C-6R$_a$, CC; 19, 3C-6R$_a$, CC; 20, 3C-6R$_a$, CC; 21, 3C-6R$_a$, CC.

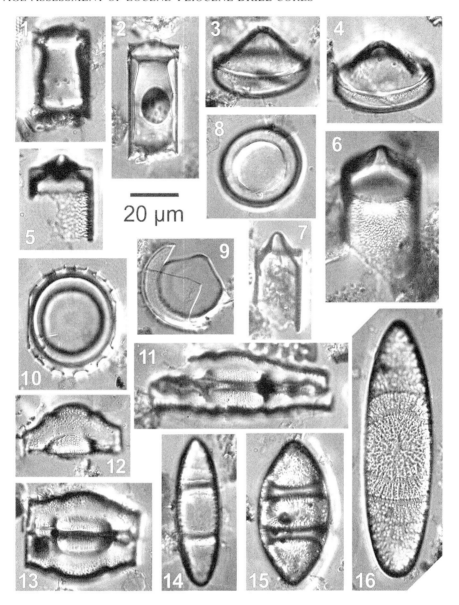

Figure 11. Light microscope images of upper Eocene diatoms from Hole 3C. Scale bar equals 20 μm. Numbers 1 and 2 show *Pseudopyxilla* sp. 1, samples: 1, 3C-6R$_a$, CC; 2, 3C-6R$_a$, CC. Numbers 3–7 show *Pseudopyxilla* sp. 2, samples: 3, 3C-6R$_a$, CC; 4, 3C-6R$_a$, CC; 5, 3C-6R$_a$, CC; 6, 3C-6R$_a$, CC; 7, 3C-6R$_a$, CC. Numbers 8 and 10 show Gen. et sp. indet. 1, samples: 8, 3C-6R$_a$, CC; 10, 3C-6R$_a$, CC. Number 9 shows *Vulcanella hannae* Sims and Mahood, sample 3C-6R$_a$, CC. Numbers 11–14 and 16 show *Goniothecium rogersii* Ehrenberg, samples: 11, 3C-6R$_a$, CC; 12, 3C-6R$_a$, CC; 13, 3C-6R$_a$, CC; 14, 3C-6R$_a$, CC; 16, 3C-6R$_a$, CC. Number 15 shows *Anaulus fossus* (?) (Grove and Sturt) Grunow, sample 3C-6R$_a$, CC.

indicative of a late Eocene-to-Oligocene age. A single pristine specimen of *Cribrocentrum reticulatum* in sample 3C-3R$_a$-1, 10 cm (7.60 mbsf) narrows the age to early late Eocene or older (≥35.9 Ma) [*Villa et al.*, 2008], assuming the specimen is not reworked. A single pristine specimen of *Chiasmolithus expansus* was also observed in sample 3C-6R$_a$-1, 87 cm (12.87 mbsf),

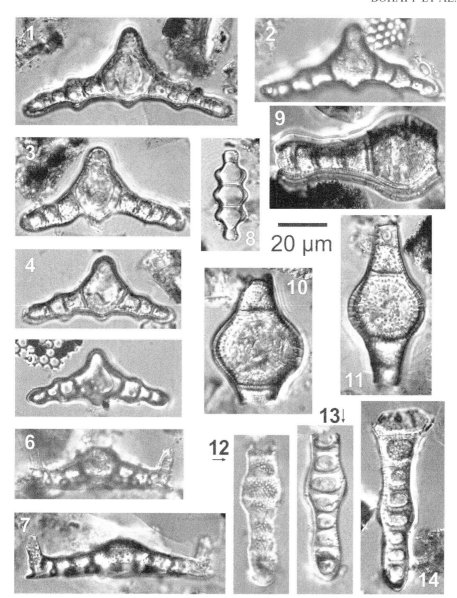

Figure 12. Light microscope images of upper Eocene diatoms from Hole 3C. Scale bar equals 20 μm. Numbers 1–7 show *Pseudorutilaria* sp. 1 (this study), samples: 1, 3C-6R$_a$, CC; 2, 3C-7R$_a$, 10 cm; 3, 3C-6R$_a$, CC; 4, 3C-6R$_a$, CC; 5, 3C-6R$_a$, CC; 6, 3C-6R$_a$, CC; 7, 3C-6R$_a$, CC. Number 8 shows *Biddulphia* sp., sample 3C-6R$_a$, CC. Numbers 9–14 show *Pseudorutilaria nodosa* Ross and Sims, samples: 9, 3C-6R$_a$, CC; 10, 3C-6R$_a$, CC; 11, 3C-6R$_a$, CC; 12, high focus, 3C-6R$_a$, CC; 13, low focus, 3C-6R$_a$, CC; 14, 3C-6R$_a$, CC.

which is characteristic of the early to mid-Eocene. The presence of these older taxa, coupled with only finding isolated, single specimens, suggests they may be reworked and therefore not reliable for dating the section.

The most robust biostratigraphic information that can be derived from the nannofossil taxa present in Hole 3C is based on the occurrence of *Transversopontis pulcheroides* (Table A3). *Wise*

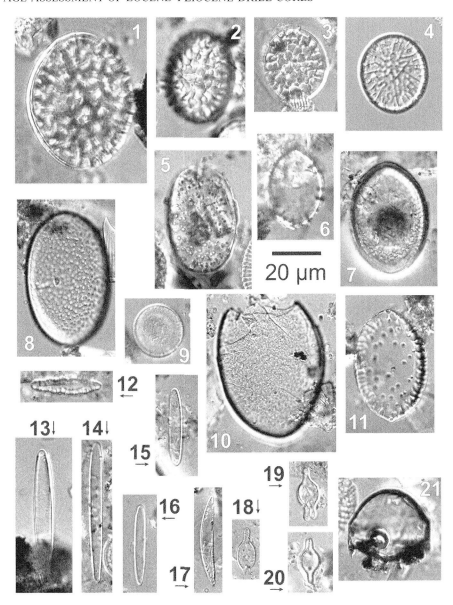

Figure 13. Light microscope images of upper Eocene diatoms from Hole 3C. Scale bar equals 20 μm. Numbers 1–4 show *Liradiscus nimbus* Suto (?), samples: 1, 3C-6R$_a$, CC; 2, 3C-6R$_a$, CC; 3, 3C-6R$_a$, CC; 4, 3C-7R$_a$, 10 cm. Number 5 shows *Xanthiopyxis* sp., sample 3C-6R$_a$, CC. Number 6 shows *Vallodiscus simplexus* Suto, sample 3C-6R$_a$, CC. Number 7 shows *Xanthiopyxis* sp., sample 3C-6R$_a$, CC. Number 8 shows *Xanthiopyxis hirsuta* Suto (?), sample 3C-6R$_a$, CC. Number 9 shows *Hyalodiscus* sp., sample 3C-7R$_a$-1, 10 cm. Number 10 shows *Xanthiopyxis* sp., sample 3C-6R$_a$, CC. Number 11 shows *Vallodiscus* sp., sample 3C-6R$_a$, CC. Number 12 shows *Vallodiscus* sp., sample 3C-6R$_a$, CC. Numbers 13 and 14 show *Ikebea* sp. B of *Scherer et al.* [2000], samples: 13, 3C-6R$_a$, CC; 14, 3C-6R$_a$, CC. Numbers 15 and 16 show small elongate spores, gen. et sp. indet., samples: 15, 3C-6R$_a$, CC; 16, 3C-6R$_a$, CC. Number 17 shows Gen. et sp. indet., sample 3C-6R$_a$, CC. Numbers 18–20 show *Kisseleviella gaster* Olney, samples: 18, 3C-6R$_a$, CC; 19, 3C-6R$_a$, CC; 20, 3C-6R$_a$, CC. Number 21 shows Gen. et sp. indet., sample 3C-6R$_a$, CC.

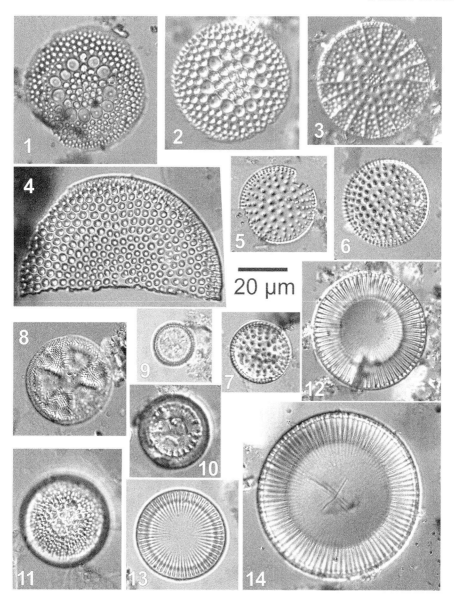

Figure 14. Light microscope images of upper Eocene diatoms from Hole 3C. Scale bar equals 20 μm. Numbers 1 and 2 show *Coscinodiscus bulliens* Schmidt, samples: 1, 3C-6R$_a$, CC; 2, 3C-6R$_a$, CC. Number 3 shows *Stictodiscus* sp., sample 3C-5R$_a$-1, 95 cm. Number 4 shows *Coscinodiscus* sp., fragment, sample 3C-6R$_a$, CC. Numbers 5–7 show *Rocella praenitida* (Fenner) Fenner, samples: 5, 3C-6R$_a$, CC; 6, 3C-6R$_a$, CC; 7, 3C-6R$_a$, CC. Number 8 shows *Actinoptychus senarius* (Ehrenberg) Ehrenberg, sample 3C-6R$_a$, CC. Number 9 shows *Distephanosira architecturalis* (Brun) Glezer, sample 3C-7R$_a$-1, 10 cm. Number 10 shows *Trochosira spinosus* Kitton, sample 3C-6R$_a$, CC. Number 11 shows *Cestodiscus convexus* Castracane, sample 3C-6R$_a$, CC. Numbers 12–14 show *Radialiplicata clavigera* (Grunow) Glezer, samples: 12, 3C-7R$_a$, CC; 13, 3C-7R$_a$-1, 10 cm; 14, 3C-6R$_a$, CC.

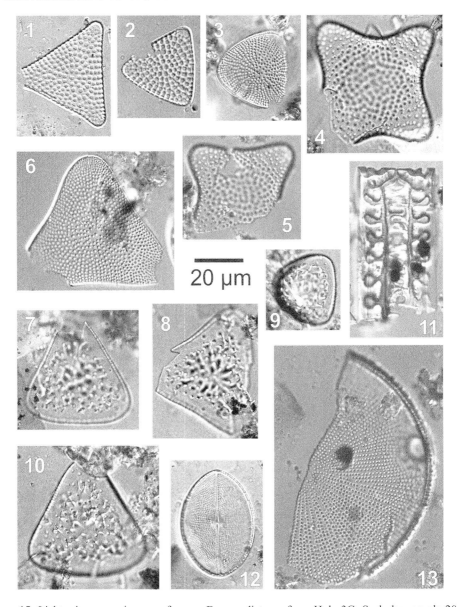

Figure 15. Light microscope images of upper Eocene diatoms from Hole 3C. Scale bar equals 20 μm. Numbers 1–10 show *Eurossia irregularis* var. *irregularis* (Greville) Sims, samples: 1, 3C-6R$_a$, CC; 2, 3C-6R$_a$, CC; 3, 3C-6R$_a$, CC; 4, quadrate form, 3C-6R$_a$, CC; 5, quadrate form, 3C-6R$_a$, CC; 6, 3C-6R$_a$, CC; 7, 3C-6R$_a$, CC; 8, 3C-6R$_a$, CC; 9, 3C-6R$_a$, CC; 10, 3C-6R$_a$, CC. Number 11 shows Gen. et sp. indet., sample 3C-6R$_a$, CC. Number 12 shows *Cocconeis* sp., sample 3C-6R$_a$, CC. Number 13 shows *Aulacodiscus* sp., sample 3C-6R$_a$, CC.

[1983] recorded the LO of *T. pulcheroides* in the lowermost Oligocene *Blackites spinosus* zone at DSDP Site 511 on the Falkland Plateau. This event was used as a regional top for correlation at that level in Cape Roberts Project Hole CRP-3 [*Watkins et al.*, 2001]. Rare to few *T. pulcheroides* specimens are present in samples from cores 3C-6R$_a$ and 3C-7R$_a$, indicating that the age at the bottom of the hole is early Oligocene or older (≥33.2 Ma, Table 3).

4.1.3. Dinoflagellate cysts. Dinoflagellate cysts are abundant throughout the Paleogene section of Hole 3C (cores 3C-1R$_a$ to 3C-7R$_a$) [*Warny and Askin*, this volume(a)]. The presence of *Deflandrea antarctica, Deflandrea cygniformis, Deflandrea phosphoritica, Enneadocysta dictyostila* (syn. *Enneadocysta partridgei*), *Spinidinium macmurdoense, Vozzhennikovia apertura,* and *Vozzhennikovia rotunda* broadly indicates an Eocene to early Oligocene age for the section (~55 to 28 Ma), based on the calibrated ranges for these taxa [*Fensome and Williams*, 2004; *Williams et al.*, 2004]. Several of these taxa, including *D. cygniformis, S. macmurdoense,* and *V. rotunda,* have LCO datums near the Eocene–Oligocene boundary in Southern Ocean sections (Table 3) [*Williams et al.*, 2004; S. Houben, personal communication, 2010], thereby constraining the age of the Hole 3C section to the Eocene. Further biostratigraphic age constraint using dinoflagellate cysts is not currently possible. Several characteristic late Eocene taxa present at the Tasman Gateway sites, such as *Deflandrea* sp. A and *Stoveracysta kakanuiensis* [*Brinkhuis et al.*, 2003a, 2003b], are not present at this site. The absence of these taxa at Site 3 may reflect environmental differences between the two different regions of the Southern Ocean or that the Hole 3C section is older than ~35.4 Ma (Subchron C16n.1n).

4.1.4. Strontium isotope dating. A $^{87}Sr/^{86}Sr$ ratio of 0.707763 was determined for a bivalve fragment from sample 3C-5R$_a$, CC (11.56 mbsf, Table 2). Using the LOWESS fit to the marine strontium isotope curve [*McArthur and Howarth*, 2004], this ratio indicates a late Eocene age of 35.9 ± 1.1 Ma for the Hole 3C section.

4.2. Holes NBP0602A-5C and NBP0602A-5D: Diatoms

Samples were prepared for diatom analysis from both holes 5C and 5D, with a primary focus on the longer section recovered in Hole 5D. Diatoms are well preserved and moderately abundant in the uppermost part of core 5D-1R$_a$ (sample 5D-1R$_a$, 50 cm; 8.5 mbsf). All taxa observed in this sample are extant species (Table A4), and the presence of *Actinocyclus actinochilus, F. curta, F. cylindrus, Fragilariopsis vanheurckii,* and *T. antarctica* indicates deposition in a sea ice-influenced environment. Although this upper interval of the hole is most likely Holocene in age, diatom biostratigraphy only constrains the age to ≤140 ka based on the absence of *R. leventerae.* This interval is assigned to the *Thalassiosira lentiginosa-Fragilariopsis kerguelensis* "b" zone of *Zielinski and Gersonde* [2002] (Figure 3).

Below the uppermost Pleistocene interval of core 5D-1R$_a$, Pliocene diatom assemblages are present within a ~7 m thick interval between sample 5D-1R$_a$-1, 95 cm and 5R$_a$-1, 25 cm (8.95 to 16.00 mbsf). The shift from Pleistocene to Pliocene diatom assemblages in Hole 5D is coincident with the boundary between lithostratigraphic units IA and IB described at ~8.80 mbsf (Shipboard Scientific Party, 2006). A similar Pliocene diatom assemblage is also noted within a soupy, "bagged" sample recovered in Hole 5C (core 5C-2R$_a$) at ~10 mbsf. Diatoms in the Pliocene intervals of holes 5C and 5D are poorly to moderately preserved and highly fragmented, but several characteristic taxa are present, including *Fragilariopsis praeinterfrigidaria, Rouxia diploneides, Thalassiosira complicata, Thalassiosira inura,* and *Thalassiosira torokina* (early form) (Table A5). The presence of *T. inura* and the absence of *Fragilariopsis barronii* place this interval of Hole 5D in the *T. inura* zone of *Gersonde and Burckle* [1990] (Figure 3).

The ages of the FO of *T. inura* and the FO of *F. barronii* have been calibrated at numerous Southern Ocean sites, although there is some discrepancy in ages between different studies. The FO of *T. inura* is placed within Subchron C3n.4n or C3r at Antarctic continental rise ODP sites 1095 and 1165, with an age of ~5.6 to 5.1 Ma [*Winter and Iwai*, 2002; *Whitehead and Bohaty,*

Table 4. Bioevents Used to Constrain the Ages of SHALDRIL II Holes 5C, 5D, 6C, and 6D

Hole(s)	Fossil Group	Present/Absent	Event	Taxon	Upper Level Present (mbsf)	Lower Level Present (mbsf)	Age[a] (Ma)	Age[b] (Ma)	Chron(s)	Age Calibration Source(s)[c]	Age Calibration Site(s)	Compilation Age (Ma)[d]	Interpreted Age Constraint (Ma)
5C-5D	diatom	absent	FO	F. barronii	-	-	4.15–4.35	4.16–4.36	C3n.1n–C3n.1r	BB, WI, WB	745, 1095, 1165	4.28–4.52	≥4.3
5C-5D	diatom	present	FO	T. complicata	8.95	16.25	5.1–5.2	5.1–5.2	C3n.4n	WI, WB	1095, 1165	4.64–4.71	≤5.1
5C-5D	diatom	present	FO	T. inura	8.95	16.25	4.9–5.55	4.9–5.6	C3n.4n–C3r	CG, WI, WB	689, 1095, 1165	4.71–4.77	≤5.6
5D	diatom	present	LO	D. praedimorpha	18.80	30.36	11.4–11.5	11.5–11.6	C5r.2n–C5r.3r	BB, HM	744, 748, 751	11.44	≥11.5
5D	diatom	present	LO	N. denticuloides	18.80	30.36	11.6–11.8	11.7–11.9	C5r.3r	GB, BB, HM, CG, MB, Bo	689, 690, 744, 747, 748	11.72	≥11.7
5D	diatom	present	FO	D. ovata	18.80	19.05	12.11	12.14	C5An.1r	CG	690	9.64–12.14	12.1
5D	diatom	present	FO	D. dimorpha v. areolata	18.80	30.36	12.2–12.74	12.2–12.8	C5Ar.1r–C5Ar.2r	BB, HM, CG	689, 690, 744, 751	12.46–12.54	≤12.8
5D	diatom	present	FO	D. praedimorpha	18.80	30.36	12.8–13.0	12.85–13.0	C5Ar.2n–C5AAn	HM, CG	689, 690, 751	12.81–13.13	≤13.1
5D	diatom	present	FCO	N. denticuloides	18.80	30.36	13.5–13.6	13.6–13.7	C5ABn–C5ABr	GB, BB, HM, CG, MB	689, 744, 747, 751	13.37–13.6	≤13.7
5D	diatom	present	FO	D. simonsenii	16.25	30.36	14.18–14.30	14.20–14.30	C5ADn	GB, BB, HM, CG, MB	689, 690, 744, 747	14.15–14.16	≤14.3

6C-6D	diatom	present	LO	*T. complicata*	5.17	17.00	2.9–3.0	2.9–3.3	C2An.1n–C2An.2r	HM, ZG	748, 1089, 1090	3.36–3.44	≥2.9
6C-6D	diatom	present	LO	*F. aurica*	10.38	17.00	3.1–3.9	3.1–3.9	C2An.1r–C2Ar	HM, ZG, WB	748, 751, 1090, 1092, 1165	3.37–4.09	≥3.1
6C-6D	diatom	absent	FO	*F. interfrigidaria*	-	-	3.8–3.9	3.8–3.9	C2Ar	GB, HM, ZG, WI, WB	690, 748, 1090, 1095, 1165	3.93–4.19	≥3.8
6C-6D	diatom	present	FO	*F. barronii*	5.17	14.99	4.15–4.35	4.16–4.36	C3n.1n–C3n.1r	BB, WI, WB	745, 1095, 1165	4.28–4.52	4.3
6C-6D	diatom	present	FO	*T. complicata*	5.17	17.00	5.1–5.2	5.1–5.2	C3n.4n	WI, WB	1095, 1165	4.64–4.71	≤5.1
6C-6D	diatom	present	FO	*T. inura*	4.58	17.00	4.9–5.55	4.9–5.6	C3n.4n–C3r	CG, WI, WB	689, 1095, 1165	4.71–4.77	≤5.6
6C	diatom	present	FO	*T. torokina* (early form)	4.58	20.30	8.5–9.0	8.6–9.1	C4Ar.2r–C4Ar.1r	BB, HM, WI	744, 748, 1095	6.43–8.03	≤9.0

[a]*Cande and Kent* [1995] and *Berggren et al.* [1995].
[b]*Gradstein et al.* [2004].
[c]Abbreviations are as follows: GB, *Gersonde and Burckle* [1990]; BB, *Baldauf and Barron* [1991]; HM, *Harwood and Maruyama* [1992]; CG, *Censarek and Gersonde* [2002]; WI, *Winter and Iwai* [2002]; ZG, *Zielinski and Gersonde* [2002]; Bo, *Bohaty et al.* [2003]; WB, *Whitehead and Bohaty* [2003a, 2003b]; and MB, *Majewski and Bohaty* [2010].
[d]Average range model. *Cody et al.* [2008].

2003a]. A considerably younger age of ~4.7 to 4.6 Ma for this event is calculated through compilation of ranges at numerous sites and statistical analysis by *Cody et al.* [2008]. Given the proximity of Site 1095 on the western side of the Antarctic Peninsula to the SHALDRIL sites, we apply an age of ~5.6 Ma for this datum, which gives a conservative estimate for the maximum age of the interval between ~9 and 16 mbsf in holes 5C and 5D. The presence of *T. complicata* supports this age interpretation and helps to further constrain the age to the Pliocene. At Site 1095, the FO of *T. complicata* is identified at ~5.1 Ma [*Winter and Iwai*, 2002], which we use as the best estimate for the maximum age of the middle section of holes 5C and 5D.

The age calibration for the FO of *F. barronii* ranges from ~4.5 to 4.2 Ma in different studies [e.g., *Baldauf and Barron*, 1991; *Winter and Iwai*, 2002; *Cody et al.*, 2008]. The younger end of this range was derived from continental rise sites 1095 and 1165 [*Winter and Iwai*, 2002; *Whitehead and Bohaty*, 2003a]. As a conservative estimate, we interpret ~4.3 Ma as the minimum age for the interval between ~9 and 16 mbsf in holes 5C and 5D. The combination of applied diatom datums constrains the depositional age of this interval between ~5.1 and 4.3 Ma (Table 4).

A major hiatus is present near the base of (or just below) core 5D-5R$_a$ at ~18 mbsf. Below this level, a ~12 m thick interval containing Miocene diatom assemblages was recovered in Hole 5D (samples 5D-6R$_a$-1, 0 cm through 13R$_a$-CC; 18.80 to 30.36 mbsf). Diatoms are moderately well preserved in most samples and rare to common in abundance in this interval (Tables A4 and A5). Characteristic diatom taxa observed throughout this lower section of Hole 5D include *Actinocyclus ingens*, *Chaetoceros* spp., *Coscinodiscus* spp., *Denticulopsis delicata*, *Denticulopsis dimorpha* var. *areolata*, *Denticulopsis praedimorpha*, *Denticulopsis simonsenii/vulgaris* group, *Fragilariopsis praecurta*, and *Thalassionema nitzschioides*. Core 5D-6R$_a$ is assigned to the *Denticulopsis ovata-Nitzschia denticuloides* zone of *Censarek and Gersonde* [2002] based on the co-occurrence of *D. ovata* and *N. denticuloides* (Table A5 and Figure 3). Although *N. denticuloides* is rare in Hole 5D samples, *D. praedimorpha* is relatively common and consistently present in cores 5D-6R$_a$ through 5D-13R$_a$ (Table A5). The LO of *D. praedimorpha* occurs near the LO of *N. denticuloides* (Table 4), thus allowing use of the LO of *D. praedimorpha* as a secondary datum to mark the top of the *D. ovata-N. denticuloides* zone and providing additional support for the zonal assignment of core 5D-6R$_a$. Below core 5D-6R$_a$, *D. ovata* was not observed. It is possible that the absence of *D. ovata* in the lowermost section of Hole 5D is due to paleoecological exclusion at this site, but *D. ovata* is not present in several samples containing moderately preserved and relatively abundant diatom assemblages (Tables A4 and A5). This suggests that the absence of *D. ovata* in this interval is of biostratigraphic significance. Therefore, we place cores 5D-7R$_a$ to 5D-13R$_a$ within the *D. dimorpha* zone, as defined by *Baldauf and Barron* [1991] and modified by *Censarek and Gersonde* [2002], based on the presence of *D. dimorpha* var. *areolata* and the absence of *D. ovata* (Table A5 and Figure 3).

Numerous middle Miocene datums have been calibrated at several sites in the Southern Ocean, and the middle Miocene interval is characterized by a relatively well-defined series of diatom bioevents, although there are some discrepancies between different studies for the ages of these events. For interpretation of the biostratigraphic ages of the Miocene section of Hole 5D, we preferentially rely on the calibrations determined from the middle Miocene section within ODP holes 689B and 690B on Maud Rise [*Gersonde and Burckle*, 1990; *Censarek and Gersonde*, 2002], but also consider the range of estimates for each event (Table 4). The Maud Rise sites are located within the Weddell Sea and are positioned at a similar latitude as the SHALDRIL II sites; thus they serve as a good reference point for Hole 5D.

The depositional age of the Miocene interval of Hole 5D (cores 5D-6R$_a$ through 5D-13R$_a$; 18.80 to 30.36 mbsf) can be constrained based on the presence of *D. dimorpha* var. *areolata*

Table 5. Summary of Age Interpretations for SHALDRIL II Holes 3C, 5C, 5D, 6C, 6D, and 12A

NBP0602A Site	Hole	Core Interval	Depth Interval (mbsf)	Age[a] (Ma)	Age Control[b]	Subepoch
3	C	3C-1R$_a$-1, 0 cm to -1R$_a$-1, 15 cm	3.00–3.15	0.14–0.00	D	late Pleistocene–Holocene
3	C	3C-1R$_a$-1, 55 cm to -7R$_a$, CC	3.55–14.12	~37–34.0	D, CN, DC, SR	late Eocene
5	C	5C-2R$_a$	~10.0	5.1–4.3	D	early Pliocene
5	D	5D-1R$_a$-1, 0 cm to -1R$_a$-1, 50 cm	8.00–8.50	0.14–0.00	D	late Pleistocene–Holocene
5	D	5D-1R$_a$-1, 95 cm to -5R$_a$-1, 25 cm	8.95–16.25	5.1–4.3	D	early Pliocene
5	D	5D-6R$_a$-1, 0 cm to -6R$_a$-1, CC	18.80–19.05	12.1–11.7	D	middle Miocene
5	D	5D-7R$_a$-1, 16 cm to -13R$_a$, CC	22.16–30.36	12.8–12.1	D	middle Miocene
6	C	6C-1R$_a$ to -6R$_a$, CC	3.90–14.99	4.3–3.8	D	early Pliocene
6	C	6C-7R$_a$-1, 100 cm to -9R$_a$, CC	17.00–20.43	5.1–4.3	D, S	early Pliocene
6	D	6D-1R$_a$, CC to -3R$_a$, CC	~5.0–10.26	4.3–3.8	D	early Pliocene
12	A	12A-1R$_{a-w}$-1, 0 cm to -1R$_{a-w}$-1, 13 cm	0.00–0.13	0.14–0.00	D	late Pleistocene–Holocene
12	A	12A-1R$_{a-w}$-1, 15 cm to -3R$_a$, CC	0.15–6.00	28.4–23.3	D, CN, SR	late Oligocene

[a]*Gradstein et al.* [2004].

[b]Abbreviations are as follows: D, diatom biostratigraphy; CN, calcareous nannofossil biostratigraphy; SR, strontium isotope dating; DC, dinoflagellate cyst biostratigraphy; S, stratigraphic position.

and *N. denticuloides*. The FO of *D. dimorpha* var. *areolata* is calibrated at ODP sites 689 and 690 at ~12.8 Ma [*Censarek and Gersonde*, 2002], although *Cody et al.* [2008] estimate a slightly younger age of 12.56–12.46 Ma for this event using their average range model. We apply the older age of ~12.8 Ma as a conservative maximum age for the Hole 5D section. The LO of *N. denticuloides* is robustly calibrated at ~11.7 Ma based on stratigraphy from several sites (Table 4) and provides a minimum age estimate for the Miocene interval of Hole 5D. Taken together, these two datums indicate a middle Miocene age of ~12.8 to 11.7 Ma for the lower section of Hole 5D.

As noted above, the apparent FO of *D. ovata* is observed within the recovered section of Hole 5D. Assuming that the absence of this taxon in the lowermost section of the hole is not environmentally controlled, application of this event allows further age division of the middle Miocene section. The FO of *D. ovata* is noted consistently above the FO of *D. dimorpha* var. *areolata* at several Southern Ocean sites, including ODP sites 747 and 748 [*Harwood and Maruyama*, 1992], ODP Site 690 [*Censarek and Gersonde*, 2002], and ODP Site 1165 (S. Bohaty, unpublished data). The age of the FO of *D. ovata*, however, is only well constrained at Site 690, which *Censarek and Gersonde* [2002] place at 12.1 Ma (Table 4). Application of this datum therefore constrains deposition of core 5D-6R$_a$ between ~12.1 and 11.7 Ma and places the lowermost interval of the hole (cores 5D-7R$_a$ and 5D-13R$_a$) between ~12.8 and 12.1 Ma (Table 4 and Figure 3). It is uncertain whether the FO datum of *D. ovata* provides a fixed age control point within the section or whether the occurrence of *D. ovata* in core 5D-6R$_a$ is indicative of the presence of a minor hiatus between cores 5D-6R$_a$ and 5D-7R$_a$.

The stratigraphic position of the major hiatus between the lower Pliocene and middle Miocene sections of Hole 5D cannot be determined precisely. The lowermost sample that contains Pliocene diatoms is sample 5D-5R$_a$, 25 cm (16.25 mbsf), and the uppermost sample that contains Miocene diatoms is sample 5D-6R$_a$, 0 cm (18.80 mbsf). In addition to a coring gap of ~2 m between cores 5D-5R$_a$ and 5D-6R$_a$, samples from the lowermost part of core 5D-5R$_a$ contain abundant silt-sized dolomite rhombs with only very rare diatom fragments; thus, no age information can be derived for these samples. Due to this uncertainty, the level of the hiatus is approximately placed at ~18 mbsf within the coring gap interval (Figure 3).

4.3. Holes NBP0602A-6C and NBP0602A-6D: Diatoms

Biostratigraphic work at Site 6 was focused on diatom analysis, due to the absence of calcareous nannofossils within cores recovered at this site. Both holes 6C and 6D were washed down to approximately 4.0 mbsf, and no core was recovered above this level (Figure 4). Samples from cores 6C-2R$_a$ to 6C-7R$_a$ (~4.0 to 17.0 mbsf) and cores 6D-1R$_a$ to 6D-3R$_a$ (~5.0 to 10.26 mbsf) contain moderately well-preserved diatoms that are few to common in abundance. Characteristic diatom taxa present in these cores include *Chaetoceros* spp. (small spores), *Coscinodiscus* spp., *Rhizosolenia* spp., *Rouxia antarctica*, *Stephanopyxis* spp., *Thalassiosira nitzschioides*, and *Thalassiothrix/Trichotoxon* spp. (Table A6). Many of the samples in this interval are very rich in diatomaceous material, but most diatom specimens are highly fragmented. Even with a high degree of fragmentation, a number of biostratigraphically useful taxa could be identified, including *D. delicata*, *F. barronii*, *F. praeinterfrigidaria*, *R. diploneides*, *T. inura*, and *T. torokina* (early form) (Table A6). Using the overlapping ranges of these taxa, a broad age of early Pliocene is interpreted for the sections recovered in holes 6C and 6D.

If the presence/absence of several diatom taxa in holes 6C and 6D is considered, the established Southern Ocean zonal scheme for the Pliocene can be applied and relatively restricted biostratigraphic ages interpreted. Cores 6C-2R$_a$ to 6C-6R$_a$ (~4.0 to 15.0 mbsf) and cores 6D-1R$_a$ to 6D-3R$_a$ (~5.0 to 10.0 mbsf) are assigned to the *F. barronii* zone, as defined by *Weaver and Gombos* [1981] and renamed by *Gersonde and Burckle* [1990] and *Zielinski and Gersonde* [2002], based on the presence of *F. barronii* and the absence of *Fragilariopsis interfrigidaria* (sensu stricto) (Table A6 and Figure 4). Core 6C-7R$_a$ (~16.0 to 17.0 mbsf) is placed in the *T. inura* zone, as defined by *Gersonde and Burckle* [1990], based on the presence of *T. inura* and the absence of *F. barronii* (Table A6 and Figure 4). *F. barronii* is generally very rare in samples from holes 6C and 6D. The absence of *F. barronii* in the lowermost section of Hole 6D may therefore be due to other factors, such as poor preservation or environmental exclusion. As such, we cannot discount the possibility that core 6C-7R$_a$ also represents deposition during the *F. barronii* zone interval.

Several diatom bioevents provide relatively precise absolute age estimates for the Pliocene section recovered in holes 6C and 6D (Table 4). The calibrated FO datum of *T. complicata* provides a maximum age estimate of ~5.1 Ma for core 6C-7R$_a$ (17.05 mbsf) and above (Table 4 and Figure 4). This age interpretation is supported by the FO datum of *T. inura*, which is assigned an age of ~5.6 Ma based on the calibration from Leg 178 cores [*Winter and Iwai*, 2002]. The age of cores 6C-2R$_a$ to 6C-6R$_a$ (~4.0 to 15.0 mbsf) and cores 6D-1R$_a$ to 6D-3R$_a$ (~5.0 to 10.0 mbsf) can be further constrained to the interval between 4.3 and 3.8 Ma, within the *F. barronii* zone, using the FO datum of *F. barronii* (4.3 Ma) and FO datum of *F. interfrigidaria* (3.8 Ma) (Table 4 and Figure 4).

Diatoms are very rare and poorly preserved in the lowermost section of Hole 6C (cores 6C-8R$_a$ and 6C-9R$_a$; ~18.0 to 20.4 mbsf). Consequently, diatom biostratigraphy does not provide a well-constrained age estimate for the lowermost section penetrated at this site, and this interval has not been given a zonal assignment. The presence of rare fragments of *D. delicata* and *T. torokina* (early form), however, constrains the age of this section to the late Miocene or early Pliocene, and the overlapping ranges of *D. delicata* and *T. torokina* provide a broad age assignment of ~9.0 to 4.0 Ma. The stratigraphic position of the section recovered in holes 6C and 6D, however, is interpreted to be ~200 m higher than the lower Pliocene section recovered in holes 5C and 5D, based on seismic stratigraphic correlations [*Smith and Anderson*, this volume]. Therefore, the age of the lowermost cores recovered in Hole 6C must be younger than ~5.1 Ma—the oldest interpreted age for the lower Pliocene section recovered in holes 5C and 5D.

4.4. Hole NBP0602A-12A

4.4.1. Diatoms. The uppermost sample examined from Hole 12A (sample 12A-1R$_a$, 8 cm; 0.08 mbsf) is barren of diatoms, but sample 12A-1R$_a$, 13 cm (0.13 mbsf) contains a moderately preserved assemblage of modern (extant) diatoms. It is unclear whether the upper ~13 cm of core 12A-1R$_a$ represents recovery of in situ sediment or downcore contamination of seafloor mud. Regardless, this uppermost diatom-bearing sample contains taxa that are characteristic of a modern sea ice assemblage, including *A. actinochilus*, *Chaetoceros* spp., *F. curta*, *F. cylindrus*, and *T. antarctica* (Table A7). This interval is presumed to be Holocene in age, but diatom biostratigraphy only constrains the age of the samples to ~140 ka or younger based on the absence of *R. leventerae* [*Zielinski et al.*, 2002]. Immediately below this level, sample 12A-1R$_a$, 15 cm (0.15 mbsf) contains a pre-Quaternary diatom assemblage, indicating that the target section was reached at very shallow depths at the site.

The distinct downcore change in diatom assemblages observed in the uppermost part of core 12A-1R$_a$ corresponds to the boundary between lithostratigraphic units I and II at 0.14 mbsf, (Table A7 and Figure 5). Below this level, the interval from sample 12A-1R$_a$, 15 cm to -3R$_a$-CC (0.15 to 6.00 mbsf) contains Oligocene–Miocene diatoms that are poorly to moderately preserved and rare to few in abundance. Fragmentation is very high in all samples examined. Common taxa observed within this interval include *Hemiaulus polycystinorum*, *Rhizosolenia hebetata* f. *hiemalis*, *R. hebetata* group, *Stephanopyxis* spp., and *Trinacria excavata* (Table A7). Fragments and whole specimens of *Stephanopyxis* spp. are particularly common in these samples. *Stephanopyxis* spp. represent a diverse group of taxa that have a neritic-planktonic habitat preference and are typical of pre-Quaternary deposits on the Antarctic shelf.

Age-diagnostic diatom taxa present between 0.15 and 6.00 mbsf in Hole 12A include *Cavitatus jouseanus*, *Cavitatus rectus*, *Kisseleviella cicatricata*, and *Kisseleviella tricoronata* (Table A7). The sensu stricto form of *C. jouseanus* commonly observed at Southern Ocean pelagic sites has a well-calibrated FO at 31.1 Ma [*Roberts et al.*, 2003], which provides a general age constraint of <~31 Ma for the Hole 12A section. In equatorial Pacific cores, the FO of *C. rectus* has a calibrated age of 28.4 Ma [*Barron et al.*, 2004]. Assuming this taxon has a similar range in the Southern Ocean, a more constrained maximum age of 28.4 Ma can be interpreted for Hole 12A (Table 3).

The two *Kisseleviella* taxa observed in Hole 12A are described from the CRP-2/2A drill core in the Ross Sea [*Olney et al.*, 2005]. In CRP-2/2A, the LO of *K. cicatricata* is recorded near the Oligocene-Miocene boundary [*Scherer et al.*, 2000; *Wilson et al.*, 2002; *Olney et al.*, 2007], which provides a minimum age of ~23.3 Ma for the strata sampled in Hole 12A. This inference is broadly supported by the presence of *K. tricoronata*, although this taxon has a slightly younger LO in the earliest Miocene (~20.9 Ma [*Wilson et al.*, 2000; *Olney et al.*, 2007]). Taken together, the combined ranges of age-diagnostic taxa present in Hole 12A indicate a late Oligocene age of ~28.4 to 23.3 Ma (Table 3). This interval corresponds to the *Azpeitia gombosi* to *Rocella gelida* zones as defined by *Harwood and Maruyama* [1992] (Figure 5), although the primary markers for these zones were not identified in Hole 12A. Representative diatom taxa from the Oligocene section of Hole 12A are illustrated in Figure 16.

4.4.2. Calcareous nannofossils. Calcareous nannofossils occur sporadically throughout Hole 12A. While some samples are barren, many contain rare to few, moderately preserved calcareous nannofossils (Table A3). The predominant taxa are reticulofenestrids, including *R. daviesii*,

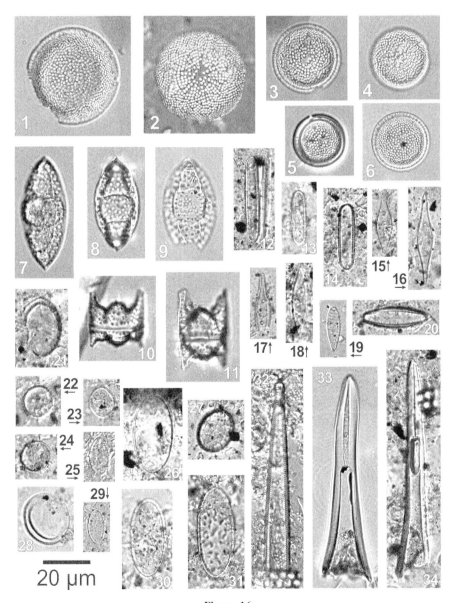

Figure 16

R. minutula, *C. floridanus*, and *Dictyococcites bisecta*. In addition, *Thoracosphaera* spp. also occur sporadically throughout the section. The limited diversity makes a precise age assignment using calcareous nannofossils difficult. The presence of *Dictyococcites bisectus*, which has a LO near the Oligocene-Miocene boundary (~22.8 Ma, Table 3) [*Poore et al.*, 1983; *Berggren et al.*, 1995; *Wilson et al.*, 2002], indicates that the section is late Oligocene or older in age.

4.4.3. Strontium isotope dating. One bivalve fragment from sample 12A-2R$_a$-2, 206 cm (4.96 mbsf) yielded a ^{87}Sr/^{86}Sr ratio of 0.708085 (Table 2). Using the LOWESS fit to the marine strontium isotope curve [*McArthur and Howarth*, 2004], this ratio indicates a late Oligocene age of 27.2±0.6 Ma for the Hole 12A section.

5. AGE SUMMARIES

In the following section, we provide a brief summary of the interpreted ages for each of the pre-Quaternary drill cores recovered at SHALDRIL II sites 3, 5, 6, and 12. Core information and diatom age/zonal assignment are summarized in Figures 2–5. All useful biostratigraphic datums are compiled in Tables 3 and 4, and a graphical summary of the age interpretations for each hole is provided in Figures 6 and 7. Diatom, calcareous nannofossil, dinoflagellate cyst, and strontium isotope age interpretations are in agreement and provide robust Eocene and Oligocene age estimates for holes 3C and 12A (Table 5). In contrast, the age interpretations for holes 5C, 5D, 6C, and 6D rely solely on diatom biostratigraphy; the presence of many well-calibrated Neogene marker taxa in these cores, however, allows reliable age determinations (Table 5).

For the biostratigraphic age interpretations presented in Figures 6 and 7, it is important to note that age ranges interpreted from biostratigraphy represent narrowest/shortest intervals that can be constrained by biostratigraphy, not the total duration/range of deposition. Given the thickness of the pre-Quaternary sections in the James Ross Basin and on the Joinville Plateau (1500–2000+ m) [*Anderson*, 1999; *Smith and Anderson*, 2010, this volume], the ~5–35 m sections sampled within each of the SHALDRIL II drill cores most likely only represent a few 100 kyrs of deposition, at most.

Figure 16. (opposite) Light microscope images of upper Oligocene diatoms from Hole 12A. Scale bar equals 20 μm. Numbers 1–6 show *Actinocyclus* sp. 1 (this study), samples: 1, 12A-1R$_{a-w}$-1, 30 cm; 2, 12A-1R$_{a-w}$-1, 30 cm; 3, 12A-1R$_{a-w}$-1, 30 cm; 4, 12A-1R$_{a-w}$-1, 30 cm; 5, 12A-1R$_{a-w}$-1, 30 cm; 6, 12A-1R$_{a-w}$-1, 30 cm. Numbers 7–11 show *Hemiaulus polycystinorum* Ehrenberg, samples: 7, oblique view, 12A-1R$_{a-w}$-1, 30 cm; 8, valve view, 12A-1R$_{a-w}$-1, 30 cm; 9, valve view, 12A-1R$_{a-w}$-1, 30 cm; 10, girdle view, 12A-1R$_{a-w}$-1, 30 cm; 11, girdle view, 12A-1R$_{a-w}$-1, 30 cm. Numbers 12–14 show *Cavitatus rectus* Akiba and Hiramatsu, samples: 12, 12A-2R$_a$-2, 220 cm; 13, 12A-1R$_{a-w}$-1, 30 cm; 14, 12A-2R$_a$, CC. Numbers 15 and 16 show *Kisseleviella cicatricata* Olney, samples: 15, 12A-2R$_a$-2, 220 cm; 16, 12A-2R$_a$-1, 150 cm. Number 17 shows *Kisseleviella* cf. *tricoronata*, sample 12A-2R$_a$-2, 220 cm. Number 18 shows *Kisseleviella tricoronata* Olney, sample 12A-2R$_a$-2, 220 cm. Number 19 shows *Kannoa hastata* Komura, sample 12A-2R$_a$-2, 220 cm. Number 20 shows *Kannoa* cf. *hastata*, sample 12A-1R$_{a-w}$-1, 96 cm. Number 21 shows *Vallodiscus simplexus* Suto, sample 12A-2R$_a$-2, 220 cm. Number 22 shows Gen. et sp. indet., sample 12A-1R$_{a-w}$-1, 30 cm. Number 23 shows *Vallodiscus* sp., sample 12A-1R$_{a-w}$-1, 30 cm. Number 24 shows *Trochosira* sp., sample 12A-2R$_a$-2, 220 cm. Numbers 25 and 29 show *Chaetoceros* sp., samples: 25, 12A-1R$_{a-w}$-1, 30 cm; 29, 12A-2R$_a$-2, 220 cm. Numbers 26, 30, and 31 show *Xanthiopyxis* spp., samples: 26, 12A-2R$_a$-1, 150 cm; 30, 12A-2R$_a$-1, 150 cm; 31, 12A-2R$_a$-2, 220 cm. Number 27 shows *Liradiscus* sp. (?), sample 12A-1R$_{a-w}$-1, 96 cm. Number 28 shows Gen. et sp. indet., sample 12A-2R$_a$-1, 150 cm. Number 32 shows *Cavitatus miocenicus* (Schrader) Akiba and Yanagisawa, sample 12A-1R$_{a-w}$-1, 30 cm. Numbers 33 and 34 show *Rhizosolenia* cf. *antarctica* Fenner, samples: 33, 12A-1R$_{a-w}$-1, 30 cm; 34, 12A-2R$_a$-2, 220 cm.

5.1. Hole NBP0602A-3C

In Hole 3C, a very thin interval containing diatom assemblages of late Pleistocene–Holocene age was recovered in the uppermost part of the section (samples 3C-1R$_a$-1, 0 cm to 3C-1R$_a$-1, 15 cm; 3.00 to 3.15 mbsf) (Tables A1 and A2). Below this level, Paleogene strata were penetrated at shallow depths, as indicated by the presence of characteristic Eocene–Oligocene age diatoms, calcareous nannofossils, and dinoflagellate cysts in the lower portion of core 3C-1R$_a$ (sample 3C-1R$_a$-1, 55 cm; 3.55 mbsf). These assemblages persist to the bottom of the hole at 14.12 mbsf (sample 3C-7R$_a$-1, CC) (Tables A1, A2, and A3 and Figure 2). Preservation and abundance of both diatoms and nannofossils improves toward the bottom of the hole, particularly in cores 3C-6R$_a$ and 3C-7R$_a$ (12.00 to 14.12 mbsf) where the flora is diverse and well preserved (Tables A1, A2, and A3). Paleogene diatoms identified in Hole 3C are dominantly neritic, tychopelagic forms that are not biostratigraphically useful, and most nannofossil and dinoflagellate cyst taxa identified in the section are long ranging. Several less abundant taxa from all three fossil groups provide general age constraints for the section. Diatom, calcareous nannofossil, and dinoflagellate cyst biostratigraphy broadly constrains the age of the Paleogene interval of Hole 3C to the late middle Eocene–early Oligocene interval (~38.5 to 28 Ma). Careful consideration of the presence/absence of several diatom and dinoflagellate cyst taxa (Table 3 and Figure 6) narrows the interpreted biostratigraphic age to the late Eocene (~37 to 34.0 Ma). One strontium isotope-based age estimate from core 3C-5R$_a$ provides support for this more restricted age interpretation, indicating an age of 35.9±1.1 Ma (Table 2 and Figure 6). Given the great thickness of the total sediment package at this site (as interpreted from seismic reflection profiles), the ~10 m thick Paleogene section sampled in Hole 3C is presumed to represent a very short interval of time within the late Eocene. Further characterization of the microfossil assemblages in Hole 3C and calibration of diatom and dinoflagellate cyst ranges in the Eocene–Oligocene boundary interval at other sites in the future may help further constrain the age of the Hole 3C section.

The upper Eocene section recovered in Hole 3C may be correlative with the uppermost La Meseta Formation, which crops out on Seymour Island [*Elliot*, 1988]. Using strontium isotope stratigraphy, *Ivany et al.* [2008] place the upper part of TELM 7 of the La Meseta Fm. within the upper Eocene. Similar strontium isotope ages are obtained for both Hole 3C (Table 2 and Figure 6) and upper TELM 7 [*Ivany et al.*, 2008] (Figure 8). Analysis of pollen and dinoflagellate cyst assemblages within Hole 3C, however, indicates a slightly younger age for Hole 3C relative to the uppermost levels of the La Meseta Formation [*Warny and Askin*, this volume(a)]; if correct, these age differences are not distinguished by strontium isotope stratigraphy. In the larger region, correlative upper Eocene sections to Hole 3C have been recovered at DSDP Site 511 on the Falkland Plateau [*Basov et al.*, 1983] and at Site 696 on the South Orkney microcontinent [*Mohr*, 1990]. Further integration of sedimentological, microfossil, and geochemical data from these sites can potentially provide greater insight into the late Eocene paleoenvironmental setting of the region.

5.2. Hole NBP0602A-12A

The uppermost interval of Hole 12A (0.0–0.13 mbsf) contains a modern diatom assemblage, biostratigraphically constrained to an age younger than ~140 ka. From 0.15 mbsf to the bottom of the hole at 6.0 mbsf (samples 12A-1R$_a$, 15 cm to 12A-3R$_a$, CC), the diatom assemblage is characterized by Oligocene taxa, including the age-diagnostic species *C. jouseanus*, *C. rectus*, *K. cicatricata*, and *K. tricoronata* (Table A7), whose combined ranges provide an age estimate of 28.4 to 23.3 Ma for the section (Table 3 and Figure 6). The Hole 12A calcareous nannofossil assemblage is of limited diversity, consisting mostly of reticulofenestrids, but the presence of

D. bisectus indicates the section is late Oligocene or older in age (≥ 22.8 Ma), thus supporting the diatom-based age interpretation. A single strontium isotope age from a bivalve shell sampled in core 12A-2R_a also supports the biostratigraphic age interpretations, indicating an age of 27.2 ± 0.6 Ma (Table 2 and Figure 6).

5.3. Holes NBP0602A-5C and NBP0602A-5D

At the present time, the only age control for holes 5C and 5D is supplied by diatom biostratigraphy. The upper part of core 5D-1R_a (sample 5D-1R_a-1, 50 cm; 8.50 mbsf) contains a late Pleistocene–Holocene assemblage, characterized by modern taxa that are typical of the Antarctic shelf sea ice zone (Tables A4 and A5). Diatom biostratigraphy constrains the age of this unit to ≤ 140 ka. A hiatus is identified at the base of core 5D-1R_a, and the interval from sample 5D-1R_a-1, 95 cm (8.95 mbsf) down through sample 5D-5R_a-1, 25 cm (16.25 mbsf) is assigned an early Pliocene age. A soupy, poorly recovered core from Hole 5C (core 5C-2R_a) is also included in this interval (Figure 3). Based on the presence of *T. complicata* and the absence of *F. barronii*, an age range of 5.1 to 4.3 Ma is interpreted for the Pliocene section of these holes (Table 4 and Figure 7). Another hiatus is identified between cores 5D-5R_a and 5D-6R_a (at ~18 mbsf), and cores 5D-6R_a through 5D-13R_a (18.80 to 30.36 mbsf) are assigned a middle Miocene age. The combined ranges of several diatom taxa constrain the age of the Miocene section to the interval between 12.8 and 11.7 Ma (Table 4). The apparent FO of *D. ovata* at ~20 mbsf may allow separation of this section into two biostratigraphic subunits (Figure 3). Applying the FO datum of *D. ovata*, core 5D-6R_a (18.80 to 19.05 mbsf) is assigned a biostratigraphic age between 12.1 and 11.7 Ma, and cores 5D-7R_a through 5D-13R_a (22.00 to 30.36 mbsf) is assigned an age between 12.8 and 12.1 Ma (Figures 3 and 7).

5.4. Holes NBP0602A-6C and NBP0602A-6D

The only age control for holes 6C and 6D is provided by diatom biostratigraphy. Diatoms are relatively common and moderately well preserved in cores 6C-2R_a to 6C-7R_a (~4.0 to 17.05 mbsf) and cores 6D-1R_a to 6D-3R_a (~5.0 to 10.0 mbsf) and contain assemblages characteristic of the early Pliocene (Table A6). Based on the presence of *T. complicata* and the absence of *F. interfrigidaria*, an age estimate of ~5.1 to 3.8 Ma is interpreted for the intervals between cores 6C-2R_a and 6C-7R_a (~4.0 to 17.05 mbsf) and cores 6D-1R_a and 6D-3R_a (~5.0 to 10.0 mbsf) (Table 4 and Figures 4 and 7). The presence of *F. barronii* in cores 6C-2R_a to 6C-6R_a (~4.0 to 15.0 mbsf) and cores 6D-1R_a to 6D-3R_a (~5.0 to 10.0 mbsf) further restricts the age of the section above ~15.0 mbsf to the interval between ~4.3 and 3.8 Ma (Table 4 and Figures 4 and 7). Low diatom abundance and poor preservation in samples from cores 6C-8R_a and 6C-9R_a precludes precise biostratigraphic age control for the lowermost section of Hole 6C (~18.0 to 20.4 mbsf). The section recovered in holes 6C and 6D, however, is interpreted to lie ~200 m stratigraphically above the lower Pliocene section recovered in holes 5C and 5D [*Smith and Anderson*, this volume]. Therefore, the age of the entire section recovered in holes 6C and 6D must be younger than ~5.1 Ma.

6. DISCUSSION

6.1. Antarctic Sea Ice History and Development of Shelf Diatom Floras

Owing to the importance of sea ice in bottom water formation, regulation of air-sea gas exchange, and the albedo of the southern high latitudes, a critical question in Cenozoic paleoclimatic history is when Antarctic sea ice first developed and the degree to which its seasonal extent and thickness has varied through time. Experiments using numerical climate models predict that sea ice did not become a prominent feature of the Antarctic margin until large ice sheets developed in East Antarctica and, hence, did not play a significant causal role in early glaciation during the Paleogene [*DeConto et al.*, 2007]. This counterintuitive prediction remains to be tested using data from the geological record.

The major obstacle to reconstructing sea ice history on the Antarctic margin is that there are very few proxies (if any) that can be used to uniquely determine the presence or absence of sea ice. Diatoms have been applied as indicators of sea ice extent and concentration within Pleistocene [e.g., *Gersonde and Zielinski*, 2000; *Smith et al.*, 2010] and Pliocene intervals [e.g., *Whitehead et al.*, 2005], but the use of diatoms for sea ice reconstructions in older intervals has been limited. Throughout the Cenozoic, both pelagic and neritic diatom floras of the Antarctic margin have experienced rapid turnover, and pre-Quaternary diatom assemblages are predominantly composed of extinct taxa. Most modern Antarctic sea ice-related taxa, such as *F. curta* and *F. cylindrus* [*Armand et al.*, 2005], do not appear in the geological record until the early Pliocene or younger intervals. Interpretation of sea ice conditions therefore requires speculation on the ecological affinities of extinct taxa. For example, a possible sea ice connection has been discussed for some members of the genus *Synedropsis* and the newly described genus *Creania* [*Olney et al.*, 2009]. This approach has been used with some success in interpreting the presence of Arctic sea ice during the middle Eocene, based on the occurrence of near monospecific laminae of *Synedropsis* spp. in conjunction with sedimentological evidence of sea ice [*Stickley et al.*, 2009]. Similar records, though, have not yet been identified at sites located on the Antarctic margin.

The diatom assemblages present in SHALDRIL II drill cores provide a coarse-resolution record of shelf conditions of the Antarctic Peninsula from the Eocene to Pliocene. No diatom taxa of possible sea ice affinity, such as *Creania lacyae* and *Synedropsis* spp. [*Olney et al.*, 2009; *Stickley et al.*, 2009], were identified in Eocene and Oligocene strata from holes 3C and 12A, respectively. Therefore, diatoms do not provide evidence for the existence of sea ice in the Eocene and Oligocene intervals at SHALDRIL II sites 3 and 12. The possible sea ice-related taxa *C. lacyae* and *Synedropsis* spp. are known from late Oligocene age strata in the Ross Sea [*Olney et al.*, 2009] but, to date, have not been observed in Oligocene or older time intervals outside of the Ross Sea.

The high abundance and diversity of neritic diatom taxa present in Hole 3C (Tables A1 and A2) likely indicate highly productive, sea ice-free conditions in the Antarctic Peninsula region during the late Eocene. A similarly diverse neritic diatom flora has been documented in Late Cretaceous-Paleocene strata from Seymour Island [*Harwood*, 1988]. Although there was major turnover in shelf diatom floras throughout the Cretaceous and Paleogene, the presence of high-diversity diatom assemblages on the Weddell Sea shelves extends back to at least the Albian (Early Cretaceous) [*Gersonde and Harwood*, 1990; *Harwood and Gersonde*, 1990].

Diatoms in the late Oligocene section recovered in Hole 12A are less diverse than the late Eocene assemblages of Hole 3C and contemporaneous Oligocene assemblages from the Ross Sea (CRP-2/2A drill core) [*Scherer et al.*, 2000; *Olney et al.*, 2007]. The lower diversity of the Hole 12A assemblages is perhaps indicative of a deeper paleowater depth for this site, which excluded taxa with a shallow, neritic paleoecological preference, and/or cooler Antarctic Peninsula shelf waters relative to the late Eocene. The Oligocene diatom assemblages of Hole 12A are also more lightly silicified and composed of smaller taxa than the Eocene assemblages of Hole 3C (e.g., compare Figures 8–15 to Figure 16), which further suggests both a deeper, pelagic setting for this site and local surface water cooling from the late Eocene to late Oligocene.

In contrast to the Eocene and Oligocene sections, diatom assemblages in the late middle Miocene and early Pliocene intervals of holes 5C, 5D, 6C, and 6D provide hints of the presence of sea ice. Extinct taxa of possible sea ice affinity, including *C. lacyae*, *F. praecurta*, *Synedropsis cheethamii*, and *Synedropsis*(?) sp. 1 (Tables A4, A5, and A6), and the modern sea ice-related taxon *F. vanheurckii* [*Gersonde and Zielinski*, 2000] occur sporadically in rare abundance throughout these cores. It is not possible to quantify the extent of sea ice cover in these Miocene and Pliocene intervals, but this evidence argues for at least the intermittent presence of seasonal sea ice. This broad interpretation is consistent with evidence of increased glacial activity from the late Oligocene (Hole 12A) to the late middle Miocene (Hole 5D) and progressive Neogene cooling of terrestrial climates in the Antarctic Peninsula [*Anderson et al.*, 2011; *Warny and Askin*, this volume(a), this volume(b); *Wellner et al.*, this volume].

6.2. Early Pliocene Warming

Many lines of evidence indicate that the early Pliocene was characterized by relative warming of Southern Ocean surface waters [*Whitehead and Bohaty*, 2003b; *Escutia et al.*, 2009], reduced sea ice concentration on the Antarctic margin [*Abelmann et al.*, 1990; *Hillenbrand and Fütterer*, 2002; *Hillenbrand and Ehrmann*, 2005; *Whitehead et al.*, 2005], and, possibly, periods of West Antarctic Ice Sheet collapse [*Naish et al.*, 2009; *Pollard and DeConto*, 2009]. In the Antarctic Peninsula region, ice caps are interpreted to have repeatedly waxed and waned through the late Miocene–early Pliocene interval and were perhaps out of phase with global glacial-interglacial periods (i.e., with maximum ice coverage corresponding to warm "interglacial" intervals) [*Bart and Anderson*, 2000; *Bart et al.*, 2005; *Smellie et al.*, 2009; *Smith and Anderson*, 2010]. Given this interpreted highly dynamic climate state, can the diatom assemblages present at Joinville Plateau sites 5 and 6 be used to infer any further paleoenvironmental information for the early Pliocene, specifically regarding temperature or sea ice conditions?

The majority of pelagic diatom taxa present in Pliocene intervals of holes 5C, 5D, 6C, and 6D are extinct (Tables A4, A5, and A6). Therefore, little information concerning paleotemperature changes can be extracted using the ecological preferences of modern taxa. Furthermore, as discussed above, the rare occurrence of several taxa with sea ice affinity most likely provides evidence of sea ice, but these observations provide no information, for example, on the extent of spring/summer sea ice. One intriguing aspect of the *F. barronii* zone (~4.3–3.8 Ma) section from holes 6C and 6D is the predominance of *Stephanopyxis* spp. Diatomaceous sands recovered in cores 6C-4R$_a$ to 6C-7R$_a$, in particular, contain very abundant diatom fragments, with common specimens of small *Stephanopyxis* spp. This diatom group is composed of several neritic-planktonic taxa that are rare in the modern Antarctic shelf sea ice flora. Thus, the abundance of this group in the Pliocene intervals of holes 6C and 6D may indicate the local presence of productive, open-water areas on the shelf during spring and summer months, with minimal sea ice cover and turbidity. Alternatively, diatom fragmentation and the high concentration of *Stephanopyxis* spp. in the sandy sediments of the Joinville Plateau could be the result of sediment focusing and winnowing which acted to substantially alter the assemblage composition. The presence of abundant small diatom fragments and numerous small taxa in most samples, however, does not support intense winnowing.

On the western side of the Antarctic Peninsula, a ~150 m thick section of lower Pliocene strata was drilled on the continental shelf at ODP Site 1097 [*Iwai and Winter*, 2002; *Bart et al.*, 2007]. At Site 1097, seismic unit S2 and the upper part of unit S3 were determined to be early Pliocene in age, but limited depositional facies and paleoenvironmental interpretations could be made for this section due to very poor core recovery [*Barker and Camerlenghi*, 2002]. Despite poor recovery, lower Pliocene diatom assemblages have been documented in this section and assigned to the *T. inura* zone (~5.6–4.3 Ma) [*Iwai and Winter*, 2002]. These assemblages are similarly characterized by an abundance of *Stephanopyxis* spp. [*Barker et al.*, 1999; *Iwai and Winter*, 2002; M. Iwai, personal communication, 2010]. This suggests the "*Stephanopyxis* assemblage" is characteristic of lower Pliocene shelf sediments of the Antarctic Peninsula region, indicating that these observations may represent a regional paleoenvironmental signal, perhaps indicative of reduced summer sea ice concentration. This interpretation is supported by increased opal (diatom) accumulation rates in continental-rise drift sediments on the western side of the Antarctic Peninsula between ~5.2 and 3.0 Ma [*Hillenbrand and Fütterer*, 2002; *Hillenbrand and Ehrmann*, 2005] and growth increment analysis of early Pliocene pectinid bivalves from Cockburn Island [*Williams et al.*, 2010], both indicating regionally reduced spring/summer sea ice cover in the early Pliocene. Evidence of greater seasonality than present day in the Antarctic Peninsula region during the early Pliocene is also interpreted from biometric analysis of fossil bryozoans obtained from deposits on James Ross Island [*Clark et al.*, 2010].

Around the East Antarctic margin, lower Pliocene sediments have been identified/obtained at several shelf sites, including the recently recovered AND-1B drill core from beneath the Ross Ice Shelf [*Naish et al.*, 2009], the Dry Valley Drilling Project (DVDP) 10, DVDP 11, and CIROS-2 drill cores in the Ross Sea [*Winter and Harwood*, 1997; *Winter et al.*, 2011b], and the Sørsdal Formation that crops out in the Vestfold Hills near Prydz Bay [*Pickard et al.*, 1988; *Whitehead et al.*, 2001]. The Sørsdal Formation is composed of sandy diatomite and diatomaceous sands deposited in a coastal setting of the East Antarctic margin. Diatom biostratigraphy places the entire formation within the *F. barronii* zone [*Harwood et al.*, 2000], and diatom evidence indicates warmer than present summer sea-surface temperatures. Deposition of diatomaceous sands of the Sørsdal Formation and the lower Pliocene section of holes 6C and 6D, therefore, are approximately contemporaneous. This suggests reduced sea ice extent around the entire Antarctic margin during the early Pliocene between ~4.3 and 3.8 Ma.

7. CONCLUSIONS

Pre-Quaternary cores ranging in age from Eocene to Pliocene were recovered by the SHAL-DRIL II project at four sites in the northern James Ross Basin and on the Joinville Plateau. The age estimates for these cores are relatively well constrained and are based primarily on diatom biostratigraphy, with supporting age information from calcareous nannofossil and dinoflagellate cyst biostratigraphy and strontium isotope chemostratigraphy. Since the recovered cores represent short/thin stratigraphic intervals of extremely thick sequences, the biostratigraphic data can only be used to interpret an age range for each core, and deposition most likely occurred during a short period within each interpreted age range. Combined biostratigraphic and strontium isotope age data indicate that the Hole 3C section is late Eocene in age (~37–34 Ma), and the Hole 12A section is late Oligocene in age (~28.4–23.3 Ma). Diatom biostratigraphy places the sequence recovered in holes 5C and 5D within the late middle Miocene (~12.8–11.7 Ma) and early Pliocene (~5.1–4.3 Ma), and an early Pliocene age (~5.1–3.8 Ma) is interpreted for cores recovered in holes 6C and 6D.

Diatom assemblages in the SHALDRIL II cores provide only limited paleoenvironmental information about shelf environments of the James Ross Basin and Joinville Plateau during the Eocene–Pliocene interval. High diversity of neritic taxa and the absence of possible sea ice taxa in upper Eocene and upper Oligocene strata of sites 3 and 12 suggest that sea ice was not a prominent feature of shelf environments of the Antarctic Peninsula in these time intervals. In contrast, rare occurrences of sea ice-related taxa are observed in middle Miocene and lower Pliocene sections recovered at sites 5 and 6, most likely indicative of long-term cooling and development of the sea ice environment between the late Oligocene and middle Miocene. A distinctive "*Stephanopyxis* assemblage" is noted in the early Pliocene age strata of sites 5 and 6. The abundance of *Stephanopyxis* in these cores may indicate limited spring/summer sea ice coverage—at least during the more prolonged interglacial intervals of the early Pliocene.

The drill cores obtained by the SHALDRIL II drilling program provide valuable snapshots of the Cenozoic paleoclimatic history of the Antarctic Peninsula region. Although long continuous records were not recovered at the targeted sites, these cores enable reconstruction of long-term climate trends and represent an important addition to available outcrop records of the region. The sediments recovered at the SHALDRIL II sites contain both terrestrial (pollen and spores) and marine microfossil assemblages that are well suited for biostratigraphic dating. The marine microfossils include opaline (diatoms), calcareous (nannofossils), and organic-walled (dinofla-gellate cysts) taxa and represent a mixture of shelf and pelagic taxa. Opaline microfossils, in particular, are well preserved, indicating that the strata have not been deeply buried. Given that studies of pre-Quaternary Antarctic shelf sediments are typically plagued by chronostratigraphic problems, the biostratigraphic and strontium isotope results from the SHALDRIL II cores demonstrate that it is possible to accurately date the thick sequences of seaward-dipping strata in the northern James Ross Basin and on the Joinville Plateau. Moreover, these cores provide a foundation for future coring programs aimed at establishing a detailed history of the paleoclimatic evolution of the Antarctic Peninsula region.

APPENDIX A: LIST OF TAXA AND TAXONOMIC NOTES

A1. CALCAREOUS NANNOFOSSILS

Braarudosphaera bigelowii (Gran and Braarud) Deflandre
Chiasmolithus expansus (Bramlette and Sullivan) Gartner
Coccolithus pelagicus (Wallich) Schiller
Coronocyclus prionion (Deflandre and Fert) Stradner and Edwards
Cribrocentrum reticulatum (Gartner and Smith) Perch-Nielsen
Cyclicargolithus cf. *floridanus* (Roth and Hay) Bukry
Cyclicargolithus floridanus (Roth and Hay) Bukry
Dictyococcites bisectus (Hay, Mohler, and Wade) Bukry and Percival
Dictyococcites productus (Kamptner) Backman
Markalius apertus Perch-Nielsen
Pontosphaera cf. *latoculata* (Bukry and Percival) Perch-Nielsen
Pontosphaera pectinata (Bramlette and Sullivan) Sherwood
Pontosphaera spp.
Reticulofenestra daviesii (Haq) Haq
Reticulofenestra filewiczii (Wise and Wiegand) Dunkley Jones
Reticulofenestra haqii Backman (3–5 μm)
Reticulofenestra minuta Roth
Reticulofenestra minutula (Gartner) Haq and Berggren
Reticulofenestra sp. (>8 μm)
Reticulofenestra sp. (5–8 μm)
Reticulofenestra spp.
Thoracosphaera heimii (Lohmann) Kamptner

Thoracosphaera spp.

Toweius pertusus (Sullivan) Romein

Transversopontis pulcheroides (Sullivan) Báldi-Beke

Watznaueria barnesae (Black) Perch-Nielsen

Watznaueria britannica (Stradner) Reinhardt

Zygrhablithus bijugatus (Deflandre) Deflandre

A2. DIATOMS

Actinocyclus actinochilus (Ehrenberg) Simonsen

Actinocyclus actinochilus (Ehrenberg) Simonsen (early form)

Notes: A variant of *Actinocyclus actinochilus* closely resembling that described and illustrated by *Harwood and Maruyama* [1992, p. 699, Plate 12, Figure 9] was observed in the lower Pliocene section of Hole 6C. These specimens are characterized by close packing of radial rows of punctae and most likely represent "early forms" of *A. actinochilus*.

Actinocyclus cf. *fryxellae* Barron

Actinocyclus ingens Rattray

Actinocyclus maccollumii Harwood and Maruyama

Actinocyclus sp. 1 (Figure 16, numbers 1–6)

Notes: This form is characteristic of the upper Oligocene diatom assemblage recovered in Hole 12A. Without analysis using scanning electron microscopy, it is unclear if this form corresponds to one of the currently described *Actinocyclus* or *Cestodiscus* taxa from the Oligocene.

Actinocyclus spp.

Actinoptychus spp. (Figure 14, number 8)

Arachnoidiscus spp.

Aulacodiscus spp. (Figure 15, number 13)

Azpeitia tabularis (Grunow) Fryxell and Sims

Anaulus spp. (Figure 11, number 15)

Biddulphia spp. (Figure 9, number 17; Figure 12, number 8)

Cavitatus jouseanus (Sheshukova-Poretzkaya) Williams

Notes: Only the sensu stricto form of *C. jouseanus* was observed in Hole 12A (see discussion of *Harwood and Bohaty* [2001]).

Cavitatus miocenicus (Schrader) Akiba and Yanagisawa (Figure 16, number 32)

Cavitatus rectus Akiba and Hiramatsu (Figure 16, numbers 12–14)

Cestodiscus convexus Castracane (Figure 14, number 11)

Chaetoceros bulbosum (Ehrenberg) Heiden

Chaetoceros spp. (Figure 16, numbers 25 and 29)

Notes: This group includes small vegetative cells and resting spores of *Chaetoceros* spp. that are typical of Neogene sediments on the Antarctic margin.

Cocconeis spp. (Figure 15, number 12)

Corethron pennatum (Grunow) Ostenfeld in Van Heurck

Coscinodiscus bulliens Schmidt (Figure 14, numbers 1 and 2)

Coscinodiscus spp. (Figure 14, number 4)

Creania lacyae Olney

Dactyliosolen antarcticus Castracane

Denticulopsis crassa Yanagisawa and Akiba

Denticulopsis delicata Yanagisawa and Akiba

Denticulopsis dimorpha var. *areolata* Yanagisawa and Akiba

Denticulopsis maccollumii Simonsen

Denticulopsis ovata (Schrader) Yanagisawa and Akiba

Denticulopsis praedimorpha Barron ex Akiba

Denticulopsis simonsenii Yanagisawa and Akiba

Denticulopsis vulgaris (Okuno) Yanagisawa and Akiba

Notes: *D. vulgaris* and *D. simonsenii* were grouped together due to difficulties in separating poorly preserved specimens of these taxa.

Distephanosira architecturalis (Brun) Glezer (Figure 14, number 9)

Notes: All specimens of *D. architecturalis* observed in Hole 3C were of small diameter (≤20 μm).

Eucampia antarctica "var. *planus*"

Notes: This "flattened" variety of *E. antarctica* was noted in the middle Miocene section of Hole 5D. It is characterized by a wide transverse length in valve view and reduced elevation height. It has also been previously recorded in the Miocene section of ODP Hole 1165B (S. Bohaty, unpublished data).

Eucampia antarctica var. *recta* (Mangin) Fryxell and Prasad

Eurossia irregularis var. *irregularis* (Greville) Sims (Figure 15, numbers 1–10)

Notes: This taxon displays a large degree of interspecific variability. In addition to three- and four-sided forms with well-developed areolation (Figure 15, numbers 1–6), hyaline forms with scattered punctae and a swollen center were also observed in Hole 3C (Figure 15, numbers 7–10). Similar forms were observed in the CRP-3 drill core [*Harwood and Bohaty*, 2001] and most likely represent hypovalves or resting spores of *E. irregularis* var. *irregularis*.

Fragilariopsis aurica (Gersonde) Gersonde and Bárcena

Fragilariopsis barronii (Gersonde) Gersonde and Bárcena

Fragilariopsis curta (Van Heurck) Hustedt

Fragilariopsis cylindrus (Grunow) Krieger

Fragilariopsis kerguelensis (O'Meara) Hustedt

Fragilariopsis obliquecostata (Van Heurck) Heiden

Fragilariopsis praecurta (Gersonde) Gersonde and Bárcena

Fragilariopsis praeinterfrigidaria (McCollum) Gersonde and Bárcena

Fragilariopsis praeinterfrigidaria-interfrigidaria (intermediate form)

Notes: Intermediate forms between *F. praeinterfrigidaria* and *F. interfrigidaria* were observed in Pliocene sediments recovered in holes 6C and 6D. These specimens possess a band of very light silicification between the trans-apical costae. Similar forms were recorded in the Pliocene section of ODP Hole 1138A [*Bohaty et al.*, 2003].

Fragilariopsis rhombica (O'Meara) Hustedt

Fragilariopsis ritscheri Hustedt

Fragilariopsis sublinearis (Van Heurck) Heiden

Fragilariopsis vanheurckii (Peragallo) Hustedt

Gen. et sp. indet. 1 (this study) (Figure 11, numbers 8 and 10)

Goniothecium rogersii Ehrenberg (Figure 11, numbers 11–14, 16)

Notes: See *Suto et al.* [2008] for a recent revision of this genus.

Hemiaulus caracteristicus Hajós (Figure 9, numbers 1–2)

Notes: Several of the *H. caracteristicus* specimens observed in Hole 3C possess a distinctive pitted pattern that mimics areolation; these specimens were separated as "*H.* cf. *caracteristicus*" (Figure 9, number 1).

Hemiaulus dissimilis Grove and Sturt (Figure 9, numbers 10–13)

Hemiaulus polycystinorum Ehrenberg (Figure 16, numbers 7–11)

Hemiaulus reflexispinosus Ross and Sims (Figure 8, numbers 1–8)

Hemiaulus sp. 1 (this study) (Figure 9, numbers 3–9)

Notes: This form is typical of the upper Eocene diatom assemblages present in Hole 3C. The valves are relatively small (≤~35 μm in basal length) with relatively thin, straight horns. Prominent straight spines are present on the horn tips of better preserved specimens. Thin hyaline ridges run between the internal bases of the horns on the mantle edge but do not extend up the side of the horns. The upper part of the mantle is coarsely areolated and divided into three or more chambers by transverse septa. The lower part of the mantle is characterized by two prominent downward extensions. Although these are spine-like features, they most likely represent support for a lightly silicified velum on the lower part of the mantle that is not preserved—a structural feature that also typifies *H. caracteristicus* [*Mahood et al.*, 1993]. *Hemiaulus* sp. 1 is similar to *Hemiaulus prolongatus* Ross and Sims of *Ross et al.* [1977], but differs in its coarse areolation of both the mantle and horns and its shorter, less prominent downward projections. *Hemiaulus* sp. 1 also has some affinity with *Hemiaulus stilwelli* Harwood and Bohaty, particularly compared to the specimen illustrated as *H. stilwelli* in Plate 3, Figure p of *Harwood and Bohaty* [2000]. *Hemiaulus* sp. 1, however, differs from *H. stilwelli* in its more robust valve structure, coarser areolation, and relatively short basal length.

Hemiaulus sp. 2 (this study)

Notes: This taxon was recognized as "*Hemiaulus* sp." by *Gombos and Ciesielski* [1983, p. 602, Plate 20, figures 7–9]. It possesses three prominent central folds (transverse sulci), small horns that curve outward, thin hyaline ridges on the internal surface of each horn, and a small, central tube that is most likely an extension of a labiate process. It is most common in the uppermost Eocene section of DSDP Hole 511, but its range also extends into the lowermost Oligocene of this hole [*Gombos and Ciesielski*, 1983].

Hemiaulus spp.

Hyalodiscus spp. (Figure 13, number 9)

Ikebea sp. B of *Scherer et al.* [2000] (Figure 13, numbers 13 and 14)

Isthmia spp.

Kannoa hastata Komura (Figure 16, number 19)

Kannoa cf. *hastata* Komura (Figure 16, number 20)

Kisseleviella tricoronata Olney (Figure 16, number 18)

Kisseleviella cf. *tricoronata* (Figure 16, number 17)

Notes: A variant of *K. tricoronata* with slightly pinched, subcapitate apices was observed in several samples from Hole 12A.

Kisseleviella cicatricata Olney (Figure 16, numbers 15 and 16)

Kisseleviella gaster Olney (Figure 13, numbers 18–20)

Large, ornamented spores (Figure 13, numbers 1–8, 10–12, and 21; Figure 16, numbers 21, 23, 26–28, and 30–31)

Notes: Many large, ornamented spores belonging to the genera *Cladogramma* Ehrenberg, *Liradiscus* Greville, *Xanthiopyxis* Ehrenberg, and *Vallodiscus* Suto, among others, were observed in holes 3C and 12A and were not taxonomically separated.

Navicula directa (Smith) Ralfs

Nitzschia denticuloides Schrader

Nitzschia grossepunctata Schrader

Nitzschia maleinterpretaria Schader

Paralia spp.

Pinnularia quadratarea (Schmidt) Cleve

"*Poretzkia*? sp." of *Hajós* [1976, Plate 17, Figure 3] (Figure 10, numbers 14–16)

Notes: This distinctive form most likely does not belong to the genus *Poretzkia* of *Jousé* [1949]. We use the designation only to maintain continuity with the work of *Hajós* [1976]. The flat, elevated portion of the valve is similar in symmetry and structure to *P. danica* (see Figure 10); therefore, this form may perhaps represent a hypovalve of *P. danica* or a separate, but related, taxon.

Porosira spp.

Porotheca danica (Grunow) Fenner (Figure 10, numbers 18–21)

Proboscia alata (Brightwell) Sundström

Pseudopyxilla americana (Ehrenberg) Forti

Pseudopyxilla sp. 1 (this study) (Figure 11, numbers 1 and 2)

Pseudopyxilla sp. 2 (this study) (Figure 11, numbers 3–7)

Pseudopyxilla spp. (Figure 10, Figures 8 and 9)

Pseudorutilaria nodosa Ross and Sims (Figure 12, numbers 9–14)

Notes: *P. nodosa* was originally described from middle-to-late Eocene cores recovered on the Falkland Plateau [*Ross and Sims*, 1987]. We observed this taxon only in the late Eocene interval of Hole 3C, primarily as broken specimens. The central chamber is the largest chamber formed by trans-apical costae and is quasi-circular when observed in valve view. Longer specimens possess an inflated area midway between the center of the valve and each apex of each arm.

Pseudorutilaria sp. 1 (this study) (Figure 12, numbers 1–7)

Notes: *Pseudorutilaria* sp. 1 is common in upper Eocene diatom assemblages recovered in Hole 3C. It is characterized by an asymmetrical, trilobate valve outline, and it may be referable to *Eunotogramma weissei* var. *producta* Grove and Sturt—a taxon described from the Oamaru Diatomite in New Zealand. This taxon is similar to *P. nodosa*, but differs in that it has an asymmetrical profile along the apical axis, coarser areolation, and an inflated central chamber. Most specimens of *Pseudorutilaria* sp. 1 also possess fewer chambers (i.e., those formed by trans-apical septa) along the valve than *P. nodosa*.

Pseudotriceratium radiosoreticulatum Grunow

Pterotheca aculeifera Grunow (Figure 10, numbers 10–13)

Pterotheca sp. 1 (this study)

Notes: This form is most likely a variety of *P. aculeifera* [see *Suto et al.*, 2009]. It is characterized by one or two large, flat keels that extend from the tip of the apical spine, similar to the specimen of *P. aculeifera* illustrated in *Suto et al.* [2009, Plate 9, Figures 37–38]. We have separated this form from *P. aculeifera* in case it proves to be of biostratigraphic utility in future studies.

Pterotheca sp. 2 (this study) (Figure 10, numbers 5 and 6)

Notes: This form is similar to *Syndendrium rugosum* [see *Suto*, 2005], except that it possesses a more cylindrical, bullet-shaped body (rather than conical) and does not possess branching processes.

Pyxilla reticulata Grove and Sturt (Figure 9, numbers 14–16)

Radialiplicata clavigera (Grunow) Glezer (Figure 14, numbers 12–14)

Rhizosolenia hebetata f. *hiemalis* Gran sensu *Harwood and Maruyama*, 1992, Plate 11, Figure 7.

Rhizosolenia cf. *antarctica* Fenner (Figure 16, numbers 33 and 34)

Notes: Compared to the sensu stricto form of *R. antarctica* [*Fenner*, 1984, p. 333, Plate 2, Figure 5], specimens identified as "*R.* cf. *antarctica*" in the upper Oligocene section of Hole 12A possess a less rounded and narrower spine tip.

Rhizosolenia sp. 1 (this study)

Notes: This taxon was illustrated by *Harwood and Maruyama* [1992, p. 703, Plate 2, Figures 3–5] as "Gen et sp. indet #3."

Rhizosolenia spp.

Rocella praenitida (Fenner) Fenner (Figure 14, numbers 5–7)

Rouxia antarctica Heiden

Rouxia diploneides Schrader

Rouxia isopolica Schrader

Rouxia naviculoides Schrader

Rhabdonema spp.

Rouxia spp.

Shionodiscus tetraoestrupii (Bóden) Alverson, Kang and Theriot

Stellarima microtrias (Ehrenberg) Hasle and Sims

Stephanogonia spp. (Figure 10, number 17)

Stephanopyxis spp.

Notes: We did not differentiate this large and diverse group of taxa. Many of the Paleogene forms of "*Stephanopyxis*" present in holes 3C and 12A most likely fall within the genera *Stephanonycites, Eustephanias, Dactylacanthis* of *Komura* [1999].

Stictodiscus spp. (Figure 14, number 3)

Syndendrium rugosum Suto (Figure 10, numbers 1–4)

Syndendrium spp. (Figure 10, number 7)

Synedropsis cheethamii Olney

Synedropsis(?) sp. 1 (this study)

Notes: This distinctive taxon was observed in the Miocene and Pliocene sections of holes 5D and 6C. It is relatively long (up to ~80 μm) and is characterized by subcapitate apices, a constricted central area, and fine uniseriate, trans-apical striae that are most prominent at the margins. This form most likely belongs to the genus *Synedropsis* Hasle or *Creania* Olney, the former of which is a modern genus with an ecological affinity to the sea ice environment. Similar specimens are illustrated by *Brady* [1979, Plate 6, Figure 6] and *Harwood and Maruyama* [1992, p. 17, Figures 6–8].

Thalassionema nitzschioides (Grunow) Mereschkowsky

Thalassiosira antarctica var. *antarctica* Comber

Thalassiosira complicata Gersonde

Thalassiosira fasciculata Harwood and Maruyama

Thalassiosira gersondei Barron–*Thalassiosira mahoodii* Barron group

Notes: *T. gersondei* Barron and *T. mahoodii* Barron are part of a large complex of middle and late Miocene *Thalassiosira* taxa present at sites near the Antarctic margin. This group requires further taxonomic study and division.

Thalassiosira gracilis var. *gracilis* (Karsten) Hustedt

Thalassiosira inura Gersonde

Thalassiosira jacksonii Koizumi and Barron

Thalassiosira oliverana (O'Meara) Makarova and Nikolaev

Thalassiosira torokina Brady (late form)

Thalassiosira torokina Brady (early form)

Notes: "Early" forms of *T. torokina* in the late Miocene–early Pliocene interval are typically large in diameter and possess a broad, flat valve face. "Late" forms of *T. torokina* present in the late Pliocene–Pleistocene are generally smaller in diameter and are more highly domed than early forms (see *Bohaty et al.* [1998] for additional discussion).

Thalassiosira tumida (Janisch) Hasle

Thalassiosira spp.

Notes: There are several small-diameter *Thalassiosira* taxa present in the Miocene and Pliocene sections of the SHALDRIL II cores which were not taxonomically separated.

Thalassiothrix/Trichotoxon spp.

Triceratium unguiculatum Greville group

"Triceratium" spp. (Figure 9, number 18)

Trinacria excavata Heiberg group

Trochosira spinosus Kitton (Figure 14, number 10)

Trochosira spp. (Figure 16, number 24)

Vulcanella hannae Sims and Mahood (Figure 11, number 9)

Xanthioisthmus panduraeformis (Pantocsek) Suto

Acknowledgments. We thank Captain Mike Watson and the crew of the RVIB *Nathaniel B. Palmer*, the Raytheon staff, and Seacore Limited (U.K.) drillers for their combined efforts, which resulted in a successful cruise. We acknowledge Lindsey Geary and Lenora Nicole Evans for shipboard laboratory assistance. Strontium isotope analysis of bivalve specimens and associated age interpretations were provided by John McArthur. Discussions with Cathy Stickley, Itsuki Suto, David Harwood, Jakub Witkowski, Masao Iwai, Sander Houben, and Peter Bijl regarding Southern Ocean biostratigraphy and diatom taxonomy were especially helpful. The manuscript benefitted from careful and constructive reviews by Cathy Stickley and David Harwood. Funding for this work was provided by the National Science Foundation, Polar Programs.

REFERENCES

Abelmann, A., R. Gersonde, and V. Spiess (1990), Pliocene-Pleistocene paleoceanography in the Weddell Sea—Siliceous microfossil evidence, in *Geological History of the Polar Oceans: Arctic Versus Antarctic, NATO ASI Ser. C*, vol. 308, edited by U. Bleil and J. Thiede, pp. 729–759, Kluwer Acad., Amsterdam, The Netherlands.

Anderson, J. B. (1999), *Antarctic Marine Geology*, 289 pp., Cambridge Univ. Press, Cambridge, U. K.

Anderson, J. B., J. S. Wellner, S. Bohaty, P. Manley, and S. W. Wise Jr. (2006), Antarctic Shallow Drilling project provides key core samples, *Eos Trans. AGU*, *87*(39), 402.

Anderson, J. B., et al. (2011), Progressive Cenozoic cooling and the demise of Antarctica's last refugium, *Proc. Natl. Acad. Sci. U. S. A.*, *108*(28), 11,356–11,360, doi:10.1073/pnas.1014885108.

Armand, L. K., X. Crosta, O. Romero, and J.-J. Pichon (2005), The biogeography of major diatom taxa in Southern Ocean sediments: 1. Sea ice related species, *Palaeogeogr. Palaeoclimatol. Palaeoecol.*, *223*, 93–126.

Baldauf, J. G., and J. A. Barron (1991), Diatom biostratigraphy: Kerguelen Plateau and Prydz Bay regions of the Southern Ocean, *Proc. Ocean Drill. Program Sci. Results*, *119*, 547–598.

Barker, P. F., and A. Camerlenghi (2002), Glacial history of the Antarctic Peninsula from Pacific margin sediments [online], *Proc. Ocean Drill. Program Sci. Results*, *178*, 40 pp. [Available at http://www-odp.tamu.edu/publications/178_SR/VOLUME/SYNTH/SYNTH.PDF.]

Barker, P. F., A. Camerlenghi, G. D. Acton, and Shipboard Scientific Party (1999), Proceedings of the Ocean Drilling Program Initial Reports, vol. 178, Ocean Drill Program, College Station, Tex. [Available at http://www-odp.tamu.edu/publications/178_IR/178TOC.HTM.]

Barron, J. A., and A. D. Mahood (1993), Exceptionally well-preserved early Oligocene diatoms from glacial sediments of Prydz Bay, East Antarctica, *Micropaleontology*, *39*(1), 29–45.

Barron, J. A., E. Fourtanier, and S. M. Bohaty (2004), Oligocene and earliest Miocene diatom biostratigraphy of ODP Leg 199 Site 1220, Equatorial Pacific [online], *Proc. Ocean Drill. Program Sci. Results*, *199*, 25 pp. [Available at http://www-odp.tamu.edu/publications/199_SR/VOLUME/CHAPTERS/204.PDF.]

Bart, P. J., and J. B. Anderson (2000), Relative temporal stability of the Antarctic ice sheets during the late Neogene based on the minimum frequency of outer shelf grounding events, *Earth Planet. Sci. Lett.*, *182*, 259–272.

Bart, P. J., D. Egan, and S. A. Warny (2005), Direct constraints on Antarctic Peninsula Ice Sheet grounding events between 5.12 and 7.94 Ma, *J. Geophys. Res.*, *110*, F04008, doi:10.1029/2004JF000254.

Bart, P. J., C. D. Hillenbrand, W. Ehrmann, M. Iwai, D. Winter, and S. A. Warny (2007), Are Antarctic Peninsula Ice Sheet grounding events manifest in sedimentary cycles on the adjacent continental rise?, *Mar. Geol.*, *236*, 1–13, doi:10.1016/j.margeo.2006.09.008.

Basov, I. A., P. F. Ciesielski, V. A. Krasheninnikov, F. M. Weaver, and S. W. Wise Jr. (1983), Biostratigraphic and paleontological synthesis: Deep Sea Drilling Project Leg 71, Falkland Plateau and Argentine Basin, *Initial Rep. Deep Sea Drill. Proj.*, *71*, 445–460.

Berggren, W. A., D. V. Kent, C. C. Swisher III, and M.-P. Aubry (1995), A revised Cenozoic geochronology and chronostratigraphy, *Spec. Publ. SEPM Soc. Sediment. Geol.*, *54*, 129–212.

Birkenmajer, K., A. Gazdzicki, K. P. Krajewski, A. Przybycin, A. Solecki, A. Tatur, and H. I. Yoon (2005), First Cenozoic glaciers in West Antarctica, *Pol. Polar Res.*, *26*(1), 3–12. [Available at http://www.polish.polar.pan.pl/ppr26/ppr26-003.pdf.]

Bohaty, S. M., R. P. Scherer, and D. M. Harwood (1998), Quaternary diatom biostratigraphy and palaeo-environments of the CRP-1 drill core, Ross Sea, Antarctica, *Terra Antart.*, *5*(3), 431–453.

Bohaty, S. M., S. W. Wise Jr., R. A. Duncan, C. L. Moore, and P. J. Wallace (2003), Neogene diatom biostratigraphy, tephra stratigraphy, and chronology of ODP Hole 1138A, Kerguelen Plateau, *Proc. Ocean Drill. Program Sci. Results*, *183*, 53 pp. [Available at http://www-odp.tamu.edu/publications/183_SR/VOLUME/CHAPTERS/016.PDF.]

Brady, H. T. (1979), The dating and interpretation of diatom zones in Dry Valley Drilling Project holes 10 and 11, Taylor Valley, South Victoria Land, Antarctica, *Mem. Natl. Inst. Polar Res.*, *13*, 150–163.

Brinkhuis, H., D. K. Munsterman, S. Sengers, A. Sluijs, J. Warnaar, and G. L. Williams (2003a), Late Eocene-Quaternary dinoflagellate cysts from ODP Site 1168, off western Tasmania, *Proc. Ocean Drill. Program Sci. Results*, *189*, 36 pp. [Available at http://www-odp.tamu.edu/publications/189_SR/105/105.htm.]

Brinkhuis, H., S. Sengers, A. Sluijs, J. Warnaar, and G. L. Williams (2003b), Latest Cretaceous-earliest Oligocene and Quaternary dinoflagellate cysts, ODP Site 1172, east Tasman Plateau, *Proc. Ocean Drill. Program Sci. Results*, *189*, 48 pp. [Available at http://www-odp.tamu.edu/publications/189_SR/106/106.htm.]

Bukry, D. (1974), Coccoliths as paleosalinity indicators: Evidence from the Black Sea, *AAPG Mem.*, *20*, 353–363.

Cande, S. C., and D. V. Kent (1992), A new geomagnetic polarity time scale for the Late Cretaceous and Cenozoic, *J. Geophys. Res.*, *97*(10), 13,917–13,951.

Cande, S. C., and D. V. Kent (1995), Revised calibration of the geomagnetic polarity timescale for the Late Cretaceous and Cenozoic, *J. Geophys. Res.*, *100*(4), 6093–6095.

Censarek, B., and R. Gersonde (2002), Miocene diatom biostratigraphy at ODP sites 689, 690, 1088, 1092 (Atlantic sector of the Southern Ocean), *Mar. Micropaleontol.*, *45*(3–4), 309–356.

Clark, N., M. Williams, B. Okamura, J. Smellie, A. Nelson, T. Knowles, P. Taylor, M. Leng, J. Zalasiewicz, and A. Haywood (2010), Early Pliocene Weddell Sea seasonality determined from bryozoans, *Stratigraphy*, *7*(2–3), 199–206.

Cody, R. D., R. H. Levy, D. M. Harwood, and P. M. Sadler (2008), Thinking outside the zone: High-resolution quantitative diatom biochronology for the Antarctic Neogene, *Palaeogeogr. Palaeoclimatol. Palaeoecol.*, *260*, 92–121, doi:10.1016/j.palaeo.2007.08.020.

DeConto, R., D. Pollard, and D. M. Harwood (2007), Sea ice feedback and Cenozoic evolution of Antarctic climate and ice sheets, *Paleoceanography*, *22*, PA3214, doi:10.1029/2006PA001350.

Desikachary, T. V., and P. M. Sreelatha (1989), *Oamaru Diatoms*, *Bibliotheca Diatomologica, Band 19*, 330 pp.

Domack, E. W., A. Leventer, S. Root, J. Ring, E. Williams, D. Carlson, E. Hirshorn, W. Wright, R. Gilbert, and G. Burr (2003), Marine sedimentary record of natural environmental and recent warming in the Antarctic

Peninsula, in *Antarctic Peninsula Climate Variability: Historical and Paleoenvironmental Perspectives*, *Antarct. Res. Ser.*, vol. 79, edited by E. Domack et al., pp. 205–224, AGU, Washington, D. C.

Edwards, A. R. (1991), The Oamaru diatomite, *N. Z. Geol. Surv. Paleontol. Bull.*, *64*, 260 pp.

Elliot, D. H. (1988), Tectonic setting and evolution of the James Ross Basin, northern Antarctic Peninsula, in *The Geology and Paleontology of Seymour Island, Antarctic Peninsula*, edited by R. M. Feldmann and M. O. Woodburne, *Mem. Geol. Soc. Am.*, *169*, 541–555.

Escutia, C., M. A. Bárcena, R. G. Lucchi, O. Romero, A. M. Ballegeer, J. J. Gonzalez, and D. M. Harwood (2009), Circum-Antarctic warming events between 4 and 3.5 Ma recorded in marine sediments from the Prydz Bay (ODP Leg 188) and the Antarctic Peninsula (ODP Leg 178) margins, *Global Planet. Change*, *69*, 170–184.

Fenner, J. (1984), Eocene-Oligocene planktic diatom stratigraphy in the low latitudes and the high southern latitudes, *Mar. Micropaleontol.*, *30*(4), 319–342.

Fenner, J. (1985), Late Cretaceous to Oligocene planktic diatoms, in *Plankton Stratigraphy*, edited by H. M. Bolli, J. B. Saunders, and K. Perch-Nielsen, pp. 713–762, Cambridge Univ. Press, Cambridge, U. K.

Fensome, R. A., and G. L. Williams (2004), *The Lentin and Williams Index of Fossil Dinoflagellates*, AASP *Contrib. Ser.*, vol. 42, 909 pp., AASP The Palynol. Soc., Notthingham, U. K.

Gersonde, R., and M. A. Bárcena (1998), Revision of the upper Pliocene-Pleistocene diatom biostratigraphy for the northern belt of the Southern Ocean, *Micropaleontology*, *44*(1), 84–98.

Gersonde, R., and D. M. Harwood (1990), Lower Cretaceous diatoms from ODP Leg 113 Site 693 (Weddell Sea), Part 1: Vegetative cells, *Proc. Ocean Drill. Program Sci. Results*, *113*, 365–402.

Gersonde, R., and U. Zielinski (2000), The reconstruction of late Quaternary Antarctic sea-ice distribution: The use of diatoms as a proxy for sea-ice, *Palaeogeogr. Palaeoclimat. Palaeoecol.*, *162*(3–4), 263–286.

Gersonde, R. E., and L. H. Burckle (1990), Neogene diatom biostratigraphy of ODP Leg 113, Weddell Sea (Antarctic Ocean), *Proc. Ocean Drill. Program Sci. Results*, *113*, 761–789.

Gombos, A. M., Jr. (1983), Middle Eocene diatoms from the South Atlantic, *Initial Rep. Deep Sea Drill. Proj.*, *71*(2), 565–581.

Gombos, A. M., Jr., and P. F. Ciesielski (1983), Late Eocene to early Miocene diatoms from the Southwest Atlantic, *Initial Rep. Deep Sea Drill. Proj.*, *71*(2), 583–634.

Gradstein, F. M., J. G. Ogg, and A. G. Smith (2004), *A Geologic Timescale 2004*, 610 pp., Cambridge Univ. Press, Cambridge, U. K.

Hajós, M. (1976), Upper Eocene and lower Oligocene diatomaceae, archaeomonadaceae, and silicoflagellatae in southwestern Pacific sediments, DSDP Leg 29, *Initial Rep. Deep Sea Drill. Proj.*, *35*, 817–884.

Hambrey, M. J., J. L. Smellie, A. E. Nelson, and J. S. Johnson (2008), Late Cenozoic glacier-volcano interaction on James Ross adjacent areas, Antarctic Peninsula region, *Geol. Soc. Am. Bull.*, *120*(5/6), 709–731, doi:10.1130/B26242.1.

Harwood, D. M. (1988), Upper Cretaceous and lower Paleocene diatoms and silicoflagellate biostratigraphy of Seymour Island, eastern Antarctic Peninsula, in *Geology and Paleontology of Seymour Island, Antarctic Peninsula*, edited by R. Feldmann and M. O. Woodburne, *Mem. Geol. Soc. Am.*, *169*, 55–129.

Harwood, D. M. (1989), Siliceous microfossils, in *Antarctic Cenozoic History from the CIROS-1 Drillhole, McMurdo Sound*, edited by P. J. Barrett, *DSIR Bull.*, *245*, 67–97.

Harwood, D. M., and S. M. Bohaty (2000), Marine diatom assemblages from Eocene and younger erratics, McMurdo Sound, Antarctica, in *Paleobiology and Paleoenvironments of Eocene Rocks, McMurdo Sound, East Antarctica*, *Antarct. Res. Ser.*, vol. 76, edited by J. D. Stilwell and R. M. Feldmann, pp. 73–98, AGU, Washington, D. C.

Harwood, D. M., and S. M. Bohaty (2001), Early Oligocene siliceous microfossil biostratigraphy of Cape Roberts Project core CRP-3, Victoria Land Basin, Antarctica, *Terra Antart.*, *8*(4), 315–338.

Harwood, D. M., and R. Gersonde (1990), Lower Cretaceous diatoms from ODP Leg 113 Site 693 (Weddell Sea), Part 2: Resting spores, chrysophycean cysts, an endoskeletal dinoflagellate, and notes on the origin of diatoms, *Proc. Ocean Drill. Program Sci. Results*, *113*, 403–425.

Harwood, D. M., and T. Maruyama (1992), Middle Eocene to Pleistocene diatom biostratigraphy of Southern Ocean sediments from the Kerguelen Plateau, Leg 120, *Proc. Ocean Drill. Program Sci. Results*, *120*, 683–733.

Harwood, D. M., A. McMinn, and P. G. Quilty (2000), Diatom biostratigraphy of the Pliocene Sørsdal Formation, Vestfold Hills, East Antarctica, *Antarct. Sci.*, *12*(4), 443–462.

Heroy, D. C., and J. B. Anderson (2007), Radiocarbon constraints on Antarctic Peninsula Ice Sheet retreat following the Last Glacial Maximum (LGM), *Quat. Sci. Rev.*, *26*, 3286–3297, doi:10.1016/j.quascirev.2007.07.012.

Hillenbrand, C. D., and W. Ehrmann (2005), Late Neogene to Quaternary environmental changes in the Antarctic Peninsula region: Evidence from drift sediments, *Global Planet. Change*, *45*, 165–191.

Hillenbrand, C. D., and D. K. Fütterer (2002), Neogene to Quaternary deposition of opal on the continental rise west of the Antarctic Peninsula, ODP Leg 178, sites 1095, 1096, and 1101 [online], *Proc. Ocean Drill. Program Sci. Results*, *178*, 33 pp. [Available at http://www-odp.tamu.edu/publications/178_SR/VOLUME/CHAPTERS/SR178_23.PDF.]

Ivany, L. C., K. C. Lohmann, F. Hasiuk, D. B. Blake, A. Glass, R. B. Aronson, and R. M. Moody (2008), Eocene climate record of a high southern latitude continental shelf: Seymour Island, Antarctica, *Geol. Soc. Am. Bull.*, *120*(5/6), 659–678, doi:10.1130/B26269.1.

Iwai, M., and D. M. Winter (2002), Data report: Taxonomic notes of Neogene diatoms from the western Antarctic Peninsula: Ocean Drilling Program Leg 178 [online], *Proc. Ocean Drill. Program Sci. Results*, *178*, 57 pp. [Available at http://www-odp.tamu.edu/publications/178_SR/VOLUME/CHAPTERS/SR178_35.PDF.]

Jousé, A. P. (1949), Algae diatomaceae aetatis supernecretaceae ex arenis argillaceis systematis fluminis Bolschoy Aktay in declivitate orientali Ural Borealis, *Not. Syst. Sect. Cryptogam. Inst. Nomine V.L. Komarovii Acad. Sci. U.R.S.S.*, *6*, 65–78.

Komura, S. (1999), Further observations on valve attachment within diatom frustules and comments on several new taxa, *Diatom*, *15*, 11–50.

Kulhanek, D. (2007), Paleocene and Maastrichtian calcareous nannofossils from clasts in Pleistocene glaciomarine muds from the northern James Ross Basin, western Weddell Sea, Antarctica, in *Antarctica: A Keystone in a Changing World—Online Proceedings of the 10th International Symposium on Antarctic Earth Sciences*, edited by A. K. Cooper, C. R. Raymond, and 10th ISAES Editorial Team, *U.S. Geol. Surv. Open File Rep., 2007-1047*, 1–5, doi:10.3133/of2007-1047.srp019.

Mahood, A. D., J. A. Barron, and P. A. Sims (1993), A study of some unusual, well-preserved Oligocene diatoms from Antarctica, *Nova Hedwigia Beih.*, *106*, 243–267.

Majewski, W., and S. M. Bohaty (2010), Surface-water cooling and salinity decrease during the middle Miocene climate transition at Southern Ocean ODP Site 747 (Kerguelen Plateau), *Mar. Micropaleontol.*, *74*(1–2), 1–14, doi:10.1016/j.marmicro.2009.10.002.

McArthur, J. M., and R. J. Howarth (2004), Strontium isotope stratigraphy, in *A Geological Timescale 2004*, edited by F. M. Gradstein, J. G. Ogg, and A. G. Smith, pp. 96–105, Cambridge Univ. Press, Cambridge, U. K.

McArthur, J. M., D. Rio, F. Massari, D. Castradori, T. R. Bailey, M. Thirlwall, and S. Houghton (2006), A revised Pliocene record for marine-$^{87}Sr/^{86}Sr$ used to date an interglacial event recorded in the Cockburn Island Formation, Antarctic Peninsula, *Palaeogeogr. Palaeoclimatol. Palaeoecol.*, *242*, 126–136.

Michalchuk, B. R., J. B. Anderson, J. S. Wellner, P. Manley, W. Majewski, and S. Bohaty (2009), Holocene climate and glacial history of the northeastern Antarctic Peninsula: The marine sedimentary record from a long SHALDRIL core, *Quat. Sci. Rev.*, *28*, 3049–3065, doi:10.1016/j.quascirev.2009.08.012.

Milliken, K. T., J. B. Anderson, J. S. Wellner, S. M. Bohaty, and P. Manley (2009), High-resolution Holocene climate record from Maxwell Bay, South Shetland Islands, Antarctica, *Geol. Soc. Am. Bull.*, *121*(11/12), 1711–1725, doi:10.1130/B26478.1.

Mohr, B. (1990), Eocene and Oligocene sporomorphs and dinoflagellate cysts from Leg 113 drill sites, Weddell Sea, Antarctica, *Proc. Ocean Drill. Program Sci. Results*, *113*, 595–612.

Mosley-Thompson, E., and L. G. Thompson (2003), Ice core paleoclimate histories from the Antarctic Peninsula: Where do we go from here?, in *Antarctic Peninsula Climate Variability: Historical and*

Paleoenvironmental Perspectives, Antarct. Res. Ser., vol. 79, edited by E. Domack et al., pp. 115–127, AGU, Washington, D. C.

Naish, T., et al. (2009), Obliquity-paced Pliocene West Antarctic ice sheet oscillations, *Nature, 458*, 322–328, doi:10.1038/nature07867.

Olney, M. P., R. P. Scherer, S. M. Bohaty, and D. M. Harwood (2005), Eocene-Oligocene paleoecology and the diatom genus *Kisseleviella* Sheshukova-Poretskaya from the Victoria Land Basin, Antarctica, *Mar. Micropaleontol., 58*(1), 56–72.

Olney, M. P., R. P. Scherer, D. M. Harwood, and S. M. Bohaty (2007), Oligocene–early Miocene Antarctic nearshore diatom biostratigraphy, *Deep Sea Res., Part II, 54*, 2325–2349, doi:10.1016/j.dsr2.2007.07.020.

Olney, M. P., S. M. Bohaty, D. M. Harwood, and R. P. Scherer (2009), *Creania lacyae* gen. nov. et sp. nov. and *Synedropsis cheethamii* sp. nov., fossil indicators of Antarctic sea ice?, *Diatom Res., 24*(2), 357–375.

Perch-Nielsen, K. (1985), Cenozoic calcareous nannofossils, in *Plankton Stratigraphy*, edited by H. Bolli, K. Perch-Nielsen, and J. B. Saunders, pp. 427–554, Cambridge Univ. Press, Cambridge, U. K.

Pickard, J., D. A. Adamson, D. M. Harwood, G. H. Miller, P. G. Quilty, and R. K. Dell (1988), Early Pliocene marine sediments, coastline, and climate of East Antarctica, *Geology, 16*, 158–161.

Pirrie, D., J. A. Crame, J. B. Riding, A. R. Butcher, and P. D. Taylor (1997), Miocene glaciomarine sedimentation in the northern Antarctic Peninsula region: The stratigraphy and sedimentology of the Hobbs Glacier Formation, James Ross Island, *Geol. Mag., 136*(6), 745–762.

Pollard, D., and R. M. DeConto (2009), Modelling West Antarctic ice sheet growth and collapse through the past five million years, *Nature, 458*, 329–332, doi:10.1038/nature07809.

Poore, R. Z., L. Tauxe, S. F. Percival Jr., J. L. LaBrecque, R. Wright, N. P. Petersen, C. C. Smith, P. Tucker, and K. J. Hsü (1983), Late Cretaceous–Cenozoic magnetostratigraphic and biostratigraphic correlations for the South Atlantic Ocean, Deep Sea Drilling Project Leg 73, *Initial Rep. Deep Sea Drill. Proj., 73*, 645–655.

Roberts, A. P., S. J. Bicknell, J. Byatt, S. M. Bohaty, F. Florindo, and D. M. Harwood (2003), Magnetostratigraphic calibration of Southern Ocean diatom datums from the Eocene-Oligocene of Kerguelen Plateau (Ocean Drilling Program sites 744 and 748), *Palaeogeogr. Palaeoclimatol. Palaeoecol., 198*, 145–168.

Ross, R., and P. A. Sims (1987), Further genera of the Biddulphiaceae (diatoms) with interlocking linking spines, *Bull. Br. Mus. Nat. Hist. Bot. Ser., 16*(4), 269–311.

Ross, R., P. A. Sims, and G. R. Hasle (1977), Observations on some species of the Hemiauloideae, *Nova Hedwigia Beih., 54*, 179–213.

Scherer, R. P., S. M. Bohaty, and D. M. Harwood (2000), Oligocene and lower Miocene siliceous microfossil biostratigraphy of Cape Roberts Project core CRP-2/2A, Victoria Land Basin, Antarctica, *Terra Antart., 7*(4–5), 417–442.

Schrader, H. J. (1976), Cenozoic planktonic diatom biostratigraphy of the southern Pacific Ocean, *Initial Rep. Deep Sea Drill. Proj., 35*, 605–671.

Sims, P. A. (1994), *Skeletonemopsis*, a new genus based on the fossil species of the genus *Skeletonema* Grev, *Diatom Res., 9*, 387–410.

Smellie, J. L., A. M. Haywood, C. D. Hillenbrand, D. J. Lunt, and P. J. Valdes (2009), Nature of the Antarctic Peninsula Ice Sheet during the Pliocene: Geological evidence and modelling results compared, *Earth Sci. Rev., 94*, 79–94, doi:10.1016/j.earscirev.2009.03.005.

Smith, J. A., C. D. Hillenbrand, C. J. Pudsey, C. S. Allen, and A. G. C. Graham (2010), The presence of polynyas in the Weddell Sea during the Last Glacial Period with implications for the reconstruction of sea-ice limits and ice sheet history, *Earth Planet. Sci. Lett., 296*(3–4), 287–298, doi:10.1016/j.epsl.2010.05.008.

Smith, R. T., and J. B. Anderson (2010), Ice-sheet evolution in James Ross Basin, Weddell Sea margin of the Antarctic Peninsula: The seismic stratigraphic record, *Geol. Soc. Am. Bull., 122*(5/6), 830–842, doi:10.1130/B26486.1.

Smith, R. T., and J. B. Anderson (2011), Seismic stratigraphy of the Joinville Plateau: Implications for regional climate evolution, in *Tectonic, Climatic, and Cryospheric Evolution of the Antarctic Peninsula*, doi:10.1029/2010SP000980, this volume.

Stickley, C. E., K. St John, N. Koc, R. W. Jordan, S. Passchier, R. B. Pearce, and L. E. Kearns (2009), Evidence for middle Eocene Arctic sea ice from diatoms and ice-rafted debris, *Nature*, *460*, 376–379, doi:10.1038/nature08163.

Strelnikova, N. I., E. Fourtanier, J. P. Kociolek, and J. A. Barron (2001), Ultrastructure studies of *Coscinodiscus* and *Cestodiscus* species from the Eocene and Oligocene, in *Lange-Bertalot-Festschrift: Studies on Diatoms*, edited by R. Jahn et al., pp. 63–96, Koeltz Sci., Koenigstein, Germany.

Suto, I. (2005), Taxonomy and biostratigraphy of the fossil marine diatom resting spore genera *Dicladia* Ehrenberg, *Monocladia* Suto and *Sydendrium* Ehrenberg in the North Pacific and Norwegain Sea, *Diatom Res.*, *20*(2), 351–374.

Suto, I., R. W. Jordan, and M. Watanabe (2008), Taxonomy of the fossil marine diatom resting spore genus *Goniothecium* Ehrenberg and its allied species, *Diatom Res.*, *23*(2), 445–469.

Suto, I., R. W. Jordan, and M. Watanabe (2009), Taxonomy of middle Eocene diatom resting spores and their allied taxa from the central Arctic Basin, *Micropaleontology*, *55*(2–3), 259–312.

Troedson, A. L., and J. L. Smellie (2002), The Polonez Cove Formation of King George Island, Antarctica: Stratigraphy, facies and implications for mid-Cenozoic cryosphere development, *Sedimentology*, *49*(2), 277–301.

Turner, J., S. R. Colwell, G. J. Marshall, T. A. Lachlan-Cope, A. M. Carleton, P. D. Jones, V. Lagun, P. A. Reid, and S. Iagovkina (2005), Antarctic climate change during the last 50 years, *Int. J. Climatol.*, *25*, 279–294, doi:10.1002/joc.1130.

Vaughan, D. G., G. J. Marshall, W. M. Connolley, C. Parkinson, R. Mulvaney, D. A. Hodgson, J. C. King, C. J. Pudsey, and J. Turner (2003), Recent rapid regional climate warming on the Antarctic Peninsula, *Clim. Change*, *60*, 243–274.

Villa, G., C. Fioroni, L. Pea, S. M. Bohaty, and D. Persico (2008), Middle Eocene–late Oligocene climate variability: Calcareous nannofossil response at Kerguelen Plateau, Site 748, *Mar. Micropaleontol.*, *69*(2), 173–192.

Warny, S., and R. Askin (2011a), Vegetation and organic-walled phytoplankton at the end of the Antarctic greenhouse world: Latest Eocene cooling events, in *Tectonic, Climatic, and Cryospheric Evolution of the Antarctic Peninsula*, doi:10.1029/2010SP000965, this volume.

Warny, S., and R. Askin (2011b), Last remnants of Cenozoic vegetation and organic-walled phytoplankton in the Antarctic Peninsula's icehouse world, in *Tectonic, Climatic, and Cryospheric Evolution of the Antarctic Peninsula*, doi:10.1029/2010SP000996, this volume.

Watkins, D. K., S. W. Wise Jr., and G. Villa (2001), Calcareous nannofossils from Cape Roberts Project Drillhole CRP-3, Victoria Land Basin, Antarctica, *Terra Antart.*, *8*(4), 339–346.

Weaver, F. M., and A. M. Gombos Jr. (1981), Southern high-latitude diatom biostratigraphy, in *The Deep Sea Drilling Project: A Decade of Progress*, edited by T. E. Warme, R. G. Douglas, and E. L. Winterer, *Spec. Publ. SEPM Soc. Sediment. Geol.*, *32*, 445–470.

Wellner, J. S., J. B. Anderson, W. Ehrmann, F. M. Weaver, A. Kirshner, D. Livsey, and A. Simms (2011), History of an evolving ice sheet as recorded in SHALDRIL cores from the northwestern Weddell Sea, Antarctica, in *Tectonic, Climatic, and Cryospheric Evolution of the Antarctic Peninsula*, doi:10.1029/2010SP001047, this volume.

Whitehead, J. M., and S. M. Bohaty (2003a), Data Report: Quaternary–Pliocene diatom biostratigraphy of ODP sites 1165 and 1166, Cooperation Sea and Prydz Bay [online], *Proc. Ocean Drill. Program Sci. Results*, *188*, 25 pp. [Available at http://www-odp.tamu.edu/publications/188_SR/VOLUME/CHAPTERS/008.PDF.]

Whitehead, J. M., and S. M. Bohaty (2003b), Pliocene summer sea surface temperature reconstruction using silicoflagellates from Southern Ocean ODP Site 1165, *Paleoceanography*, *18*(3), 1075, doi:10.1029/2002PA000829.

Whitehead, J. M., P. G. Quilty, D. M. Harwood, and A. McMinn (2001), Early Pliocene paleoenvironment of the Sørsdal Formation, Vestfold Hills, based on diatom data, *Mar. Micropaleontol.*, *41*(41), 125–152.

Whitehead, J. M., S. Wotherspoon, and S. M. Bohaty (2005), Minimal Antarctic sea ice during the Pliocene, *Geology*, *33*(2), 137–140.

Williams, G. L., H. Brinkhuis, M. A. Pearce, R. A. Fensome, and J. W. Weegink (2004), Southern Ocean and global dinoflagellate cyst events compared: Index events for the Late Cretaceous–Neogene [online], *Proc. Ocean Drill. Program Sci. Results*, *189*, 98 pp. [Available at http://www-odp.tamu.edu/publications/189_SR/VOLUME/CHAPTERS/107.PDF.]

Williams, M., et al. (2010), Sea ice extent and seasonality for the early Pliocene northern Weddell Sea, *Palaeogeogr. Palaeoclimatol. Palaeoecol.*, *292*, 306–318, doi:10.1016/j.palaeo.2010.04.003.

Wilson, G. S., et al. (2000), Chronostratigraphy of CRP-2/2A, Victoria Land Basin, Antarctica, *Terra Antart.*, *7*(4–5), 647–654.

Wilson, G. S., et al. (2002), Integrated chronostratigraphic calibration of the Oligocene-Miocene boundary at 24.0 ± 0.1 Ma from the CRP-2A drill core, Ross Sea, Antarctica, *Geology*, *30*(11), 1043–1046.

Winter, D., C. Sjunneskog, R. Scherer, P. Maffioli, C. Riesselman, and D. Harwood (2011a), Pliocene–Pleistocene diatom biostratigraphy of nearshore Antarctica from the AND-1B drill core, McMurdo Sound, *Global Planet. Change*, doi:10.1016/j.gloplacha.2010.04.004, in press.

Winter, D., C. Sjunneskog, R. Scherer, P. Maffioli, and D. Harwood (2011b), Diatom-based correlation of early to mid-Pliocene drill cores from the southwestern Ross Sea, Antarctica, *Global Planet. Change*, doi:10.1016/j.gloplacha.2009.12.004, in press.

Winter, D. M., and D. M. Harwood (1997), Integrated diatom biostratigraphy of late Neogene drillholes in Southern Victoria Land and correlation to Southern Ocean records, in *The Antarctic Region: Geological Evolution and Processes*, edited by C. A. Ricci, pp. 985–992, Terra Antart. Publ., Siena, Italy.

Winter, D. M., and M. Iwai (2002), Data report: Neogene diatom biostratigraphy, Antarctic Peninsula Pacific margin, ODP Leg 178 rise sites [online], *Proc. Ocean Drill. Program Sci. Results*, *178*, 24 pp. [Available at http://www-odp.tamu.edu/publications/178_SR/VOLUME/CHAPTERS/SR178_29.PDF.]

Wise, S. W., Jr. (1983), Mesozoic and Cenozoic calcareous nannofossils recovered by Deep Sea Drilling Project Leg 71 in the Falkland Plateau region, Southwest Atlantic Ocean, *Initial Rep. Deep Sea Drill. Proj.*, *71*(2), 481–550.

Zielinski, U., and R. Gersonde (2002), Plio-Pleistocene diatom biostratigraphy from ODP Leg 177, Atlantic sector of the Southern Ocean, *Mar. Micropaleontol.*, *45*(3–4), 225–268.

Zielinski, U., C. Bianchi, R. Gersonde, and M. Kunz-Pirrung (2002), Last occurrence datums of the diatoms *Rouxia leventerae* and *Rouxia constricta*: Indicators for marine isotope stages 6 and 8 in Southern Ocean sediments, *Mar. Micropaleontol.*, *46*(1–2), 127–137.

S. M. Bohaty, School of Ocean and Earth Science, University of Southampton, National Oceanography Centre, European Way, Southampton SO14 3ZH, UK. (S.Bohaty@noc.soton.ac.uk)

K. Jemison and S. W. Wise Jr., Department of Earth, Ocean and Atmospheric Sciences, Florida State University, 108 Carraway Building, Tallahassee, FL 32306-4100, USA.

D. K. Kulhanek, Department of Paleontology, GNS Science, PO Box 30368, Lower Hutt 5040, New Zealand.

C. Sjunneskog, Antarctic Research Facility, 108 Carraway Building, Florida State University, Tallahassee, FL 32306-4100, USA.

S. Warny, Department of Geology and Geophysics and Museum of Natural Science, Louisiana State University, E235 Howe-Russell Bldg., Baton Rouge, LA 70803, USA.

Magnetic Properties of Oligocene-Eocene Cores From SHALDRIL II, Antarctica

Luigi Jovane[1]

Department of Geology, Western Washington University, Bellingham, Washington, USA

Kenneth L. Verosub

Department of Geology, University of California, Davis, California, USA

The past climate of Antarctica is generally recognized as an important component for understanding modern climate processes. One of the most significant events in the Antarctic climate record is the Oi-1 climate change (33.55 Ma), which marked the onset of continent-wide glaciation in Antarctica at the beginning of the Oligocene and which turned the world 3°C–4°C colder in a single and abrupt event. Outcropping sedimentary sections of this time period are not globally abundant; moreover, only a few marine cores are available, and sites around the Antarctic continent from this period are especially lacking. We studied the magnetic properties of two cores collected from the Antarctic Peninsula in 2006 by the research vessel icebreaker *Nathaniel B. Palmer* during the SHALDRIL II expedition. The cores provide information about variations in magnetic properties during the late Eocene and late Oligocene. Mineral magnetic parameters show that the mineralogy and grain size of the magnetic carriers did not vary significantly between the Eocene and the Oligocene. However, the concentration of magnetic material is higher in the Oligocene section than in the Eocene section. This increase in the delivery of magnetic material into the basin is probably due to an increased sediment flux caused by icebergs or other ice-related processes. Small-amplitude variations in the magnetic parameters of the Oligocene section suggest a periodically forced process driving the flux of magnetic particles.

---

[1]Now at Instituto Oceanográfico da Universidade de São Paulo, São Paulo, Brazil.

Tectonic, Climatic, and Cryospheric Evolution of the Antarctic Peninsula
Special Publication 063
10.1029/2011SP001100

1. INTRODUCTION

Reconstruction of the past climate in Antarctica is increasingly seen as relevant for understanding changes in the present global climate. A persistent but oscillatory ice sheet (dry-based ice regime) was present in Antarctica after the late Miocene-early Pliocene [*Escutia et al.*, 2005], or ~3 Ma [*Rebesco et al.*, 2006], but previously, glaciers were intermittent and fluctuating (wet-based ice regime). In order to gain a broader perspective on large-scale climatic change, we need to look further back in time to the middle Eocene when Antarctica was only slightly covered with ice [*Zachos et al.*, 2001].

Between the middle Eocene and the early Oligocene, a fundamental change took place in the Earth's climate and oceanic circulation system. This change is the greenhouse-icehouse transition, and it turned the deep-sea waters colder by 3°C–4°C [*Zachos et al.*, 1996] in a single and abrupt climatic event that lasted less than a hundred thousand years: the Oi-1 event (~34–33.5 Ma) [*Zachos et al.*, 2001, *Zachos and Kump*, 2005]. Before this event, the conditions in the Eocene were characterized by a long-term cooling trend from the Paleocene with no permanent ice at the poles [*Pearson et al.*, 2001]. The Eocene was interspersed with short-term events recognized in numerous marine proxies that indicate unstable paleoceoanographic conditions prior to the main climatic transition [*Merico et al.*, 2008]. In addition, two episodes of significant glaciation were recognized from a magnetic study of the CIROS-1 core [*Barrett et al.*, 1989; *Wilson et al.*, 1998].

The Eocene-Oligocene transition is marked by a rapid step in $\delta^{18}O$ of over 1‰ [*Coxall et al.*, 2005; *Lear et al.*, 2004], reflecting ice growth in Antarctica and cooling of the deep-sea waters. By the Early Oligocene (Rupelian), the Antarctic ice sheet was 25% of today's dimensions [*Katz et al.*, 2008].

The main aim of the Shaldril II project was to recover Cenozoic cores from the vicinity of the Antarctic Peninsula (Figure 1) in order to reconstruct the tectonic and climatic evolution of Antarctica and the opening of the Drake Passage [*Anderson*, 1999]. For this purpose, the research vessel icebreaker *Nathaniel B. Palmer* sailed during the austral summer of the 2006 in the northern part of Weddell Sea to the tip of the Antarctic Peninsula (Joinville and Seymour islands) collecting cores from the Joinville Plateau and the James Ross Basin. The biostratigraphy of the recovered sediments indicated that they were Holocene, Plio-Pleistocene, mid-Miocene, late Oligocene, and late Eocene in age.

In this chapter, we report on the magnetic properties of two cores from the Weddell Sea side of the Peninsula (NBP0602A-3C and NBP0602A-12A), which sampled part of the late Oligocene and part of the late Eocene. Our results provide information about changes in the magnetic properties of the cores, which are related to differences in detritus delivered to the sites during these times.

2. AGES OF SAMPLED SEDIMENT

Core NBP0602A-3C was drilled at the northern edge of the James Ross Basin (63°50.861'S, 54°39.207'W) in a water depth of 340 m. The core bottomed out at 20 mbsf, but there was no recovery from the last 3 m. Of the remaining 17 m, only 6.31 m of sediment was recovered, which is divided into seven subcores (total average recovery 37%). The recovery was highest (65%–79%) between 7.5 and 13.5 mbsf. The sediment from this core mostly consists of poorly sorted, muddy, silty, very fine to fine sand with a few subangular to subrounded pebbles [*Wellner et al.*, this volume].

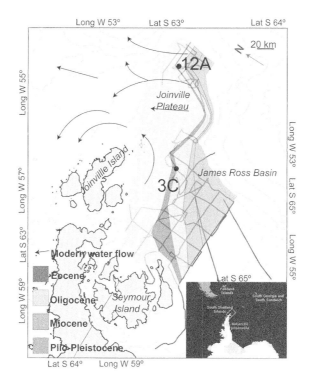

Figure 1. Locations of SHALDRIL (NBP06-02-12A as 12A and NBP06-02-3C as 3C) cores on a map of the Antarctic Peninsula sector, modified from the work of *Anderson et al.* [2006]. Arrows depict modern ocean currents from the work of *Ardelan et al.* [2010]. Shaded areas with various tonalities represent ages of strata from the work of *Anderson et al.* [1992].

Core NBP0602A-12A was drilled on the Joinville Plateau (63°16.354′S, 52°49.501′W) in a water depth of 442 m. The coring reached 7.2 mbsf, but the cored section was only 4.3 m long. Only 2.74 m was recovered as three subcores (total recovery 64%). The sediment from this hole consists of sandy mud and muddy sand. The section has been divided into six lithologic units mostly formed by blackish sandy mud with few coarse pebbles and some clay lenses [*Wellner et al.*, this volume].

The top tens of centimeters of core NBP0602A-3C contain well-preserved modern diatoms (Holocene) typical of a sea ice assemblage. Below this, diatoms become rare and poorly preserved and indicate a late Eocene age. Toward the bottom of the core, there is increased preservation and abundance of diatoms that yield an age of ~37 to 34 Ma (late Eocene) [*Bohaty et al.*, this volume]. A shell sample at the base of the NBP0602A-3C-5 core yielded a ^{87}Sr/^{86}Sr age of 35.9 Ma [*Bohaty et al.*, this volume]. At 12.35 mbsf, a sandstone cobble larger than the core barrel prevented further core recovery. In summary, the diatom biostratigraphy and a single radiometric date indicate a late Eocene age for the older strata sampled at Hole NBP0602A-3C.

In Hole NBP0602A-12A, the uppermost sediment (0–0.13 mbsf) also contains a modern diatom assemblage. The diatom assemblage from 0.15 to 7.2 mbsf includes rare to few late Oligocene taxa, including some age-diagnostic species (*Cavitatus jouseanus, Cavitatus rectus, Kisseleviella cicatricata*, and *Kisseleviella tricoronata*) (~28.4–23.3 Ma) for the section [*Anderson et al.*, 2006]. The calcareous nannofossil assemblage is not very diverse, consisting mostly of reticulofenestrids, but supporting the late Oligocene (~23.8–24.2 Ma) age [*Bohaty et al.*, this volume].

3. SAMPLES METHODOLOGY

Open-sided 2 × 2 cm square U-channels with a length of up to 1.5 m were used to sample the core, which is housed at the Antarctic Marine Geology Research Facility in Tallahassee, Florida [*Nagy and Valet*, 1993; *Weeks et al.*, 1993]. Samples were collected parallel to the long axis of sediment cores in order to obtain high-quality and high-resolution data, with measurements are done every centimeter and progressive demagnetization done in steps of 5 or 10 mT. Owing to edge effects, the first and the last four measurements of a U-channel are not considered reliable and were thus discarded. The procedure involved pushing an inverted U-channel into the split face core followed by extraction of the sediments using a nylon string. During this operation, some gaps were produced, and some of these were filled with foam. Core NBP0602A-3C was sampled with seven U-channels, and we noted four unfilled gaps and six pieces of foam. Core NBP0602A-12A was sampled with three U-channels, and the gaps occurred mainly in the first 60 cm. These circumstances reduce the number of usable measurements and contribute to the

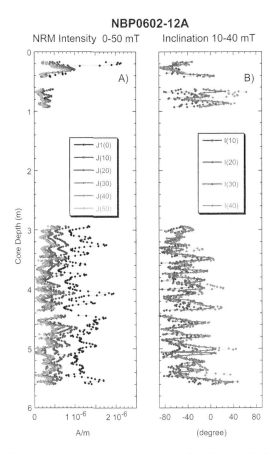

Figure 2. (a) Intensity of the natural remanent magnetization for U-channels from core NBP06-02-12A demagnetized from 0 to 50 mT. Most of the intensity is lost during the first demagnetization step, and the last steps do not always show a decrease in intensity. (b) Inclination of the natural remanent magnetization demagnetized from 0 to 40 mT. Most of inclinations have high negative values, corresponding to normal polarity. While there are some values that are positive, there are no intervals that consistently show inclinations greater than zero. Cyclicity in the lower section of the core appears to be present.

discontinuity of the magnetic record, which makes it more difficult to produce a coherent magnetostratigraphy. Furthermore, between the U-channels, there are gaps. So despite the fact that the SHALDRIL U-channels consist of soft and recently collected sediment, they are not in suitable condition for detailed magnetostratigraphic study.

Henceforth, core NBP0602A-3C and core NBP0602A-12A are referred to as core 3C and core 12A, respectively.

We measured both natural remanences (NRM) and laboratory-induced remanences and their demagnetizations. The measurements of the U-channels were made using a 2G Enterprises automated cryogenic magnetometer (Model 755). In-line alternating field (AF) demagnetization was applied in steps of 0, 5, 10, 15, 20, 25, 30, 35, 40, 50, and 60 mT. After alternating field demagnetization of the NRM, the magnetic properties were studied by imparting laboratory-induced magnetizations to the U-channels and then demagnetizing them. These measurements allowed us to study variations in the composition, concentration and grain size of the magnetic minerals.

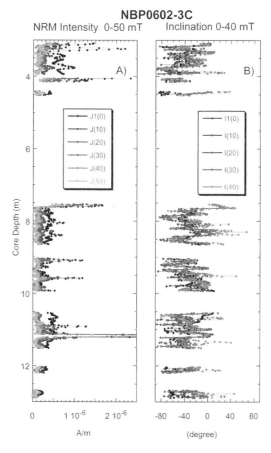

Figure 3. (a) Intensity of the natural remanent magnetization for U-channel from core NBP06-02-3C demagnetized from 0 to 50 mT. Most of the intensity is lost during the first demagnetization step, and the last steps do not always show a decrease in intensity. (b) Inclination of the natural remanent magnetization demagnetized from 0 to 40 mT. Most of inclinations have high negative values, corresponding to normal polarity. While there are some values that are positive, there are no intervals that consistently show inclinations greater than zero. No magnetostratigraphic interpretation is possible.

The induced magnetizations comprised anhysteretic remanent magnetization (ARM) acquired in a 100 mT alternating field with a superimposed 50 μT direct-field bias, isothermal remanent magnetization (IRM) at 900 mT and backfield IRM (BIRM) at 100 mT (BIRM at 100) and 300 mT (BIRM at 300). Derived parameters, such as S-ratio (S-ratio$_{300}$ = BIRM at 300/IRM) and hard IRM (HIRM) (HIRM$_{300}$ = (IRM + BIRM at 300)/2) were used to investigate the coercivity of the magnetic minerals [*Stoner et al.*, 1996]. Also, the ratio of ARM to IRM was used to determine magnetic grain size (ARM/IRM) [e.g., *King et al.*, 1982; *Opdyke and Channell*, 1996; *Stoner et al.*, 1996]. In addition, we measured hysteresis properties and first-order reversal curves (FORC) on selected samples [*Pike et al.*, 1999; *Roberts et al.*, 2000]. Saturation magnetization (M_s), saturation remanence (M_r), and coercive force (H_c) were determined from hysteresis loops, and remanent coercivity (H_{cr}) from back-field measurements with an alternating gradient magnetometer (MicroMag Model 3900) at Western Washington University. The peak field used to determine M_r from the hysteresis loops corresponds to the saturation field of the IRM (0.9 T).

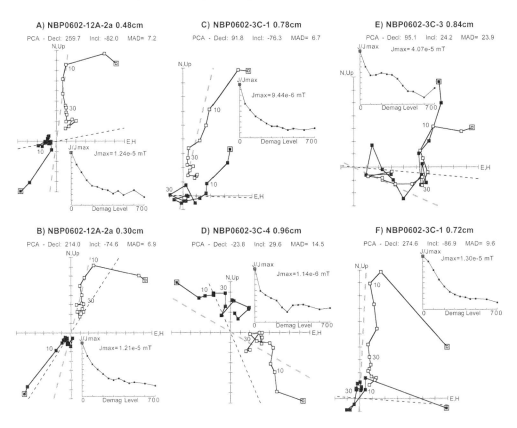

Figure 4. Orthogonal vector component diagrams and intensity decay curves during alternating field demagnetization of selected samples from NBP06-02-12A and Hole NBP06-02-3C. Solid (open) squares denote the projection on the horizontal (vertical) plane. The demagnetization levels are represented in mT. (a and b) Samples are from core NBP06-02-12A; there is a subhorizontal soft component of magnetization, which is removed after 10 mT. Even though these are the samples with the best vector component diagrams and intensity decay, the MAD angle calculated with principal component analysis (PCA) analysis [*Kirschvink*, 1980] has very high values. (c–f) Similarly, samples with the best vector component and intensity decay from core NBP06-02-3C are shown. Samples in Figures 4d and 4e may be of reverse polarity, but all of them have high MAD angles. Gray and black dashed lines are the PCA calculations for inclination and declination, respectively.

A plot of M_r/M_s versus H_{cr}/H_c was used to obtain information about magnetic domain state and magnetic grain sizes [*Day et al.*, 1977; *Dunlop*, 2002]. Thermomagnetic runs up to 700°C was performed with a Kappabridge KLY-4 (noise level 2×10^{-8} SI), on selected samples, using a furnace (CS3). The FORC data were used to study microcoercivity distributions and magnetic interactions. Reduction of the FORC data to produce FORC diagrams was done using the FORCIT software [*Acton et al.*, 2007] and FORCinel software [*Harrison and Feinberg*, 2008]. Spectral analyses were done using Analyserie 2.0 software package [*Paillard et al.*, 1996].

Electric resistivity and magnetic susceptibility, which were measured with a multisensor core logger, were reported by *Anderson et al.* [2006]. Electric resistivity was measured using a noncontact coil array (Middlebury horizontal probes and electrodes in the Wenner configuration), and magnetic susceptibility was measured using a loop (Bartington Meter MS2C). The core was logged for these parameters at 2 cm intervals.

4. NATURAL REMANENT MAGNETIZATION

For core 3C, the NRM ranges from 1.13×10^{-5} to 2.43×10^{-8} A m^{-1} with an arithmetic mean of 6.11×10^{-7}, and for core 12A, the NRM ranges from 2.13×10^{-6} to 1.50×10^{-7} A m^{-1} with an arithmetic mean of 1.04×10^{-6} (Figures 2 and 3, respectively). Owing to the high latitude, the expected inclination should be almost vertical and consequently easy to discern. Unfortunately,

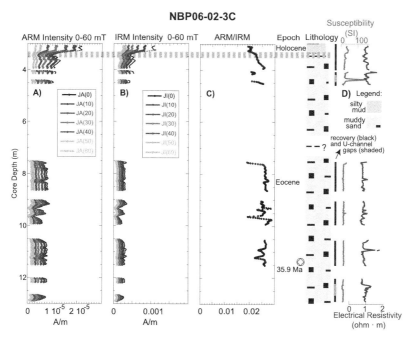

Figure 5. Mineral magnetic parameters and their demagnetization between 0 and 60 mT for core NBP06-02-3C as a function of stratigraphic position. (a) Demagnetization of anhysteretic remanent magnetization (ARM), (b) demagnetization of isothermal remanent magnetization (IRM), and (c) ARM/IRM at 0.9 T. Higher values of this interparametric ratio indicate relatively higher concentrations of finer particles. ARM and IRM demagnetizations show that the magnetic concentrations and magnetic mineralogy remains constant along most of both cores. The diamond symbol shows the 35.9 Ma strontium dating at the base of core NBP06-02-3C-5. (d) Lithology, recovery (black line), U-channel gaps (light gray line), magnetic susceptibility, and electrical resistivity are also shown [*Anderson et al.*, 2006].

only a small number of samples display linear decay of NRM on orthogonal vector plots during progressive demagnetization (Figure 4). Most of the samples show complex decay, with multiple magnetizations and mainly normal (positive) inclinations. For core 12A, we recognize a soft overprint that is removed with alternating field demagnetization at a level between 10 and 15 mT. This low-coercivity overprint is probably a viscous magnetization acquired during the coring, storage, and/or handling phases. Owing to the complexity of the demagnetization behavior, we are not able to determine the direction of the characteristic remanent magnetization (ChRM) using principal component analysis and so cannot divide the record into normal and reversed magnetozones.

5. RESULTS

For core 3C, the ARM ranges from 2.19×10^{-5} to 2.06×10^{-6} A m^{-1} with an arithmetic mean of 8.49×10^{-6}, and the IRM ranges from 1.11×10^{-3} to 8.17×10^{-5} A m^{-1} with an arithmetic mean of 3.50×10^{-4} (Figure 5). For both of these parameters, the concentration of magnetic material is about twice as high in the top 20 cm than in the rest of the core. Below the top 20 cm, the values decrease for another 20 cm and then remain generally uniform with only minor variations. Plots of the demagnetized and normalized demagnetized values of the ARM and IRM also show only small down-core variations (Figure 5). The fact that the down-core changes are minor indicates that the variations in mineral magnetic properties (ARM, IRM, and the ratios) are minimal in this interval.

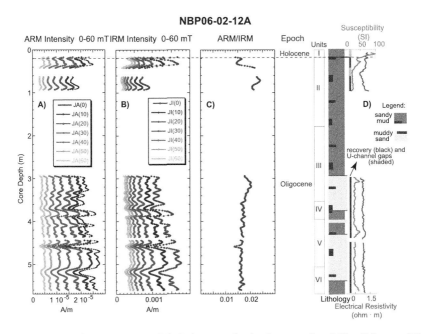

Figure 6. Mineral magnetic parameters and their demagnetization between 0 and 60 mT for core NBP06-02-12A as a function of stratigraphic position. (a) Demagnetization of ARM, (b) demagnetization of IRM, and (c) ARM/IRM at 0.9 T are shown. Higher values of this interparametric ratio indicate relatively higher concentrations of finer particles. ARM and IRM demagnetizations show that the magnetic concentrations and magnetic mineralogy remain constant along most of both cores. (d) Lithology, recovery (black line), U-channel gaps (light shaded line), magnetic susceptibility, and electrical resistivity are also shown [*Anderson et al.*, 2006].

For core 12A, the ARM ranges from 2.77×10^{-5} to 1.46×10^{-5} A m^{-1} with an arithmetic mean of 2.22×10^{-5}, and the IRM ranges from 1.75×10^{-3} to 6.61×10^{-4} A m^{-1} arithmetic mean of 1.29×10^{-3}. These parameters show a similar trend, which indicates that the concentration of magnetic material is about twice as high in the first 14 cm than in the rest of the core, where it is of generally uniform concentration (Figure 6). The S-ratio$_{300}$ for core 12A has values that range between 0.94 and 1, which is indicative of magnetite [*Stoner et al.*, 1996]. HIRM$_{300}$ has a value of approximately 2.3×10^{-3} A m^{-1} in the first 30 cm and then decreases to about 6.0×10^{-4} A m^{-1}. The values then increase down core, reaching a maximum of 1.6×10^{-3} A m^{-1} (Figure 7).

Hysteretic curves and back-field analysis on selected samples from the upper part of 3C and the lower part of 12A demonstrate that similar magnetic carriers are present in both cores and that these carriers represent a low coercivity magnetic mineral (Figure 8).

Thermomagnetic curves (Figure 9) show that, upon heating, there is a slight increase in low-field magnetic susceptibility until ~550°C, where the susceptibility shows a distinct peak. This peak is due to alteration. This is confirmed from the thermomagnetic curves run under a constant flux of argon of the same samples. After this peak, there is a major drop in susceptibility at 560°C–580°C (Figure 9), which is consistent with the presence of magnetite as main magnetic carrier [e.g., *Hunt et al.*, 1995]. During cooling, the magnetic susceptibility increased significantly, which is consistent with thermochemical alteration of iron-bearing minerals (clays or pyrite). However, there is a narrow peak just below the Curie temperature point for sample NBP06-02-12A-1 at 30 cm (Figures 9b and 9c), which is the Hopkinson peak for magnetite [*Hopkinson*, 1889].

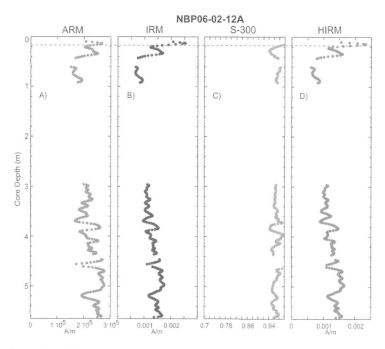

Figure 7. Stratigraphic variation in magnetic properties for NBP06-02-12A including (a) ARM, (b) IRM, (c) S-ratio$_{300}$ = backfield IRM (BIRM) at 300/IRM at 900, and (d) HIRM$_{300}$ = (IRM at 900 + BIRM at 300)/2. The S-ratio$_{300}$ and the HIRM$_{300}$ parameters show that the magnetic mineral is magnetite and that there are no variations along the core.

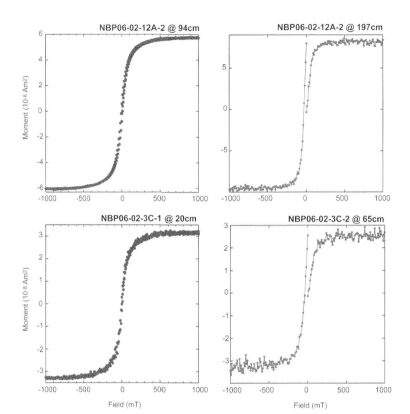

Figure 8. Hysteresis curves for a sample from core 12A, section 2, at the depth of 94 cm and from core 3C, section 1, at the depth of 20 cm. Back field curves and M_s acquisition curves for samples from core 12A, section 2, at the depth of 197 cm and from core 3C, section 2, at the depth of 65 cm are shown.

FORC diagrams of the same samples show a single peak near the origin, which is consistent with a low coercivity mineral as the main magnetic carrier (Figure 10). No magnetic interactions are indicated from the FORC diagrams. All of this information supports the conclusion that the magnetic carrier is magnetite. Given this, we use the ratios between the coercivities (H_{cr}/H_c) and the magnetizations (M_r/M_s) to determine that the magnetic domain state falls near the boundary between pseudo-single-domain (PSD) grains and multidomain (MD) grains (Figure 11). The same results are found by *Brachfeld and Banarjee* [2000] for core PD92-30 from the Palmer Deep in Andvord Bay on the Pacific side of the Antarctic Peninsula. Interestingly, the samples from core 12A (late Oligocene) fall closer to the PSD field, while the samples from core 3C (late Eocene) fall closer to the MD field.

For the lower section of core 12A (Unit II-III), a spectral analysis in depth domain was carried out (Figure 12) in order to search for astronomical modulation. We applied the Tukey window, with a removed linear trend, unit variance, and prewhitening with a bandwidth of 1.05. The power spectrum results to have significant peaks at ~220, ~25, and ~11 cm.

6. DISCUSSION

The Holocene layers at the top centimeters of both holes have almost twice the concentration of magnetic material as the late Oligocene and late Eocene sediments in these cores. This probably

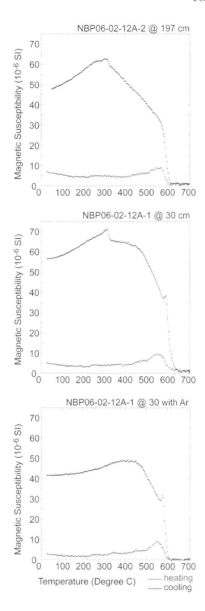

Figure 9. Thermomagnetic curves for samples (top) NBP06-02-12A-2 at 197 cm in air and NBP06-02-12A-1 (middle) at 30 in air and (bottom) in argon. Low-field magnetic susceptibility is shown for increasing (shaded) and decreasing (black) temperatures.

results from more intense glacial erosion of magnetic source rocks during the late Pleistocene and Holocene. Otherwise, the magnetic properties of the Holocene sediments and the late Eocene and late Oligocene sediments are virtually the same.

The ARM/IRM graphs show values that remain fairly uniform for each hole. However, ARM/IRM for core 12A is about 50% higher than it is for core 3C, indicating that the material is slightly finer-grained in core 12A than in core 3C [*King et al.*, 1982] (Figures 5 and 6). This conclusion is

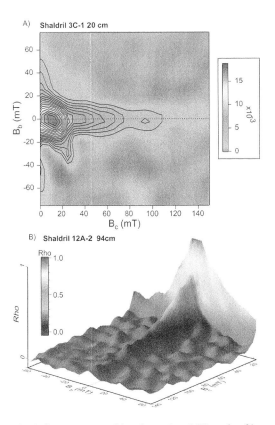

Figure 10. Representative FORC diagrams smoothing factor 5 and (Gaussian filter of 36 mT) for (a) core 3C, section 1, and (b) core 12A, section 2, the same samples for which hysteresis curves are presented. The dominant peak near the origin is consistent with the presence of magnetite. The fact that there are no other peaks indicates the exclusion of the presence of other high-coercivity mineral.

reliable because magnetite is the principal magnetic mineral, and there are no interactions [*Yamazaki and Ioka*, 1997]. Samples from core 3C (with lower IRM and ARM) are characterized by hysteresis parameters that plot well within the MD region of a "Day Plot" [*Day et al.*, 1977], which corresponds to grain sizes larger than 10 μm (MD) [*Dunlop*, 2002] (Figure 11). Samples from core 12A (with lower IRM and ARM) are characterized by hysteresis parameters that span the boundary between the PSD region and the MD region. Magnetic particles of the samples that fall in the PSD region correspond to grain sizes of 1 and 10 μm for magnetite and titanomagnetite [*Day et al.*, 1977] (Figure 11). We note, however, that only a small number of samples have been used for this analysis, and these might not be representative of the entire cores.

Core 3C from the James Ross Basin (in front of Joinville Island) is more proximal to the Antarctic continent than core 12A on the Joinville Plateau (Figure 1). Plots of NRM, ARM, and IRM intensity show clearly that the concentration of magnetic minerals in core 12A (late Oligocene) is greater than in core 3C (late Eocene), despite the close proximity of Site 3C to volcanic sources in and near James Ross Island. We attribute this to delivery of magnetic particles to Site 12A by contour currents, which have had a strong influence on sedimentation in this area [see *Wellner et al.*, this volume].

We found in the lower sections of core 12A the peaks of periodicity at ~220, ~25, and ~11 cm. The ratio ~220/~25 (8.8) is comparable to the ratio between long-term eccentricity and obliquity

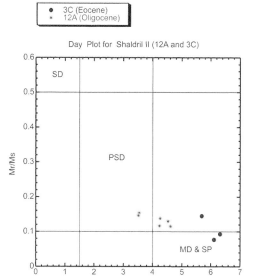

Figure 11. "Day plot" of samples from core 3C (solid circle) and from core 12A (shaded circle) [*Day et al.*, 1977]. Samples that fall in the multidomain (MD) region correspond to grain sizes larger than 10 μm [*Dunlop*, 2002]. Samples that fall in the pseudo-single-domain (PSD) region correspond to grain sizes of 1 to 10 μm for magnetite [*Day et al.*, 1977; *Brachfeld and Banerjee*, 2000].

(400/41 = 9.7). The ratio ~220/~11 (20.0) is similar to the ratio between long-term eccentricity and precession (400/21 = 19.0). The ratio ~25/~11 (2.27) is analogous to the ratio between obliquity and precession (41/21 = 1.95). For further information regarding astronomical signals in geological proxies and their uncertainties see *Paillard* [2001], *Hinnov* [2004], and *Laskar et al.* [2004]. Because we cannot estimate the rates of sedimentation and given the poor age

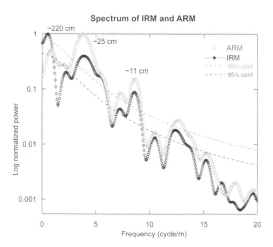

Figure 12. Spectral analysis on ARM and IRM in the depth domain using Tukey window, with bandwidth of 1.05 and error estimation on the power spectrum between 0.475843 and 5.3622. Confidence interval factor is represented by dashed lines. Spectral analysis was done using Analyseries 2.0 software package [*Paillard et al.*, 1996].

constraints, we cannot draw further conclusions. Nevertheless, the presence of this cyclicity suggests that during the late Oligocene, the sedimentation was influenced by short-term fluctuations in the magnetic properties, and this was likely controlled by contour currents.

Environmental magnetism and mineral magnetic analysis establishes that the magnetic mineralogy of the two cores is very similar. However, core 3C has lower magnetic intensity and coarser magnetic grain size than core 12A. One way to explain these differences is to postulate that between the late Eocene and late Oligocene, there was a major climatic event, likely associated with the greenhouse-icehouse transition. Although the Eocene-Oligocene Boundary was not sampled in the SHALDRIL cores, it is interesting to compare variations in the magnetic properties of the late Eocene and late Oligocene sediments. Specifically, the sediments of core 3C do not show any variations that can be related to changes in provenance or source of the magnetic particles. This is consistent with seismic evidence that the source of these sediments was confined to the James Ross and Seymour Island area [*Smith and Anderson*, 2010]. In contrast, sediments of core 12A display variations in magnetic character that is consistent with the influence of contour currents on sedimentation in this area.

7. CONCLUSION

The magnetic properties from NBP06-02-3C and NBP06-02-12A remain nearly constant during the time intervals represented by these cores, late Eocene and late Oligocene, respectively. The main magnetic carrier is uniformly magnetite. However, the magnetic concentration in core NBP06-02-12A is greater. It is also finer and PSD, while the magnetic material of core NBP06-02-3C is less abundant, coarser, and MD. For both holes, the abundance of magnetic material is highest in the younger glacimarine sediments in the tops of these cores. The greater concentration of magnetic material in late Oligocene sediments of Site 12A is attributed to the contribution from distant sources by contour currents that flow across this location. The coarser magnetic particles at Site 3C are consistent with the proximity of this site to volcanic sources at James Ross Island and more localized sediment dispersal of these sediments from their source.

Acknowledgments. We are grateful to John Anderson, Woody Wise, Julia Smith Wellner, Matt Olney, and the AMGRF Committee for allowing us to use these cores. We thank the personnel of Antarctic Marine Geology Research Facility in Tallahassee for preparing the U-channels and M. Burak Yikilmaz for preparing the MicroMag samples at UC Davis. We appreciate the corrections and suggestions from the anonymous reviewers and the editors John Anderson and Julia Smith Wellner. We also acknowledge Pat Manley of Middlebury College who performed the onboard measurements.

REFERENCES

Acton, G., A. Roth, and K. L. Verosub (2007), Analyzing micromagnetic properties with FORCIT software, *Eos Trans. AGU*, *88*, 230.

Anderson, J. B. (1999), *Antarctic Marine Geology*, 289 pp., Cambridge Univ. Press, Cambridge, U. K.

Anderson, J. B., S. S. Shipp, and F. P. Siringan (1992), Preliminary seismic stratigraphy of the northwestern Weddell Sea continental shelf, in *Recent Progress in Antarctic Earth Science*, edited by Y. Yoshida et al., pp. 603–612, Terra Sci., Tokyo.

Anderson, J. B., J. S. Wellner, S. Bohaty, P. L. Manley, and S. W. Wise Jr. (2006), Antarctic Shallow Drilling Project provides key core samples, *Eos Trans. AGU*, *87*(39), 402.

Ardelan, M. V., O. Holm-Hansen, C. D. Hewes, C. S. Reiss, N. S. Silva, H. Dulaiova, E. Steinnes, and E. Sakshaug (2010), Natural iron enrichment around the Antarctic Peninsula in the Southern Ocean, *Biogeosciences*, *7*, 11–25.

Barrett, P. J., M. J. Hambrey, D. M. Harwood, A. R. Pyne, and P. N. Webb (1989), Synthesis, *Antarctic Cenozoic History from the CIROS-1 Drillhole, McMurdo Sound, Antarctica*, edited by P. J. Barrett, *Dep. Sci. Ind. Res. Bull. 245*, 241–251, DSIR Publ., Wellington, New Zealand.

Bohaty, S. M., D. K. Kulhanek, S. W. Wise Jr., K. Jemison, S. Warny, and C. Sjunneskog (2011), Age assessment of Eocene–Pliocene drill cores recovered during the SHALDRIL II expedition, Antarctic Peninsula, in *Tectonic, Climatic, and Cryospheric Evolution of the Antarctic Peninsula*, doi:10.1029/2010SP001049, this volume.

Brachfeld, S. A., and S. K. Banerjee (2000), Rock-magnetic carriers of century-scale susceptibility cycles in glacial-marine sediments from the Palmer Deep, Antarctic Peninsula, *Earth Planet. Sci. Lett.*, *176*, 443–455.

Coxall, H. K., P. A. Wilson, H. Palike, C. H. Lear, and J. Backman (2005), Rapid stepwise onset of Antarctic glaciation and deeper calcite compensation in the Pacific Ocean, *Nature*, *433*, 53–57.

Day, R., M. D. Fuller, and V. A. Schmidt (1977), Hysteresis properties of titanomagnetites: Grain size and composition dependence, *Phys. Earth Planet. Inter.*, *13*, 260–266.

Dunlop, D. J. (2002), Theory and application of the Day plot (M_{rs}/M_s versus H_{cr}/H_c): 2. Application to data for rocks, sediments, and soils, *J. Geophys. Res.*, *107*(B3), 2057, doi:10.1029/2001JB000487.

Escutia, C., L. De Santis, F. Donda, R. B. Dunbar, A. K. Cooper, G. Brancolini, and S. L. Eittreim (2005), Cenozoic ice sheet history from East Antarctic Wilkes Land continental margin sediments, *Global Planet. Change*, *45*(1–3), 51–81, doi:10.1016/j.gloplacha.2004.09.010.

Harrison, R. J., and J. M. Feinberg (2008), FORCinel: An improved algorithm for calculating first-order reversal curve distributions using locally weighted regression smoothing, *Geochem. Geophys. Geosyst.*, *9*, Q05016, doi:10.1029/2008GC001987.

Hinnov, L. (2004), Earth's orbital parameters and cycle stratigraphy, in *A Geologic Time Scale*, edited by F. M. Gradstein, J. G. Ogg, and A. G. Smith, pp. 55–62, Cambridge Univ. Press, Cambridge, U. K.

Hopkinson, J. (1889), Magnetic and other physical properties of iron at high temperature, *Philos. Trans. R. Soc. London Ser. A*, *180*, 443–465.

Hunt, C., B. M. Moskowitz, and S. K. Banerjee (1995), Magnetic properties of rocks and minerals, in *Rock Physics & Phase Relations: A Handbook of Physical Constants*, AGU Ref. Shelf, vol. 3, edited by T. J. Ahrens, pp. 189–204, doi:10.1029/RF003p0189, AGU, Washington, D. C.

Katz, M. E., K. G. Miller, J. D. Wright, B. S. Wade, J. V. Browning, B. S. Cramer, and Y. Rosenthal (2008), Stepwise transition from the Eocene greenhouse to the Oligocene icehouse, *Nat. Geosci.*, *1*, 329–334, doi:10.1038/ngeo179.

King, J., S. K. Banerjee, J. Marvin, and O. Ozdemir (1982), A comparison of different magnetic methods for determining the relative grain size of magnetite in natural materials: Some results from lake sediments, *Earth Planet. Sci. Lett.*, *59*, 404–419.

Kirschvink, J. L. (1980), The least-squares line and plane and the analysis of paleomagnetic data, *Geophys. J. R. Astron. Soc.*, *62*, 699–718.

Laskar, J., P. Robutel, F. Joutel, M. Gastineau, A. Correia, and B. Levrard (2004), A long-term numerical solution for the insolation quantities of the Earth, *Astron. Astrophys.*, *428*, 261–285.

Lear, C. H., Y. Rosenthal, H. K. Coxall, and P. A. Wilson (2004), Late Eocene to early Miocene ice sheet dynamics and the global carbon cycle, *Paleoceanography*, *19*, PA4015, doi:10.1029/2004PA001039.

Merico, A., T. Tyrell, and P. A. Wilson (2008), Eocene/Oligocene ocean deacidification linked to Antarctic glaciation by sea-level fall, *Nature*, *452*, 979–982.

Nagy, E. A., and J.-P. Valet (1993), New advances for paleomagnetic studies of sediment cores using U-channels, *Geophys. Res. Lett.*, *20*, 671–674.

Opdyke, N. D., and J. E. T. Channell (Eds.) (1996), *Magnetic Stratigraphy*, Int. Geophys. Ser., vol. 64, edited by R. Dmowska and J. R. Holton, 346 pp., Academic, Burlington, Mass.

Paillard, D. (2001), Glacial cycles: Toward a new paradigm, *Rev. Geophys.*, *39*(3), 325–346.

Paillard, D., L. Labeyrie, and P. Yiou (1996), Macintosh program performs time-series analysis, *Eos Trans. AGU*, *77*, 379.

Pearson, P. N., P. W. Ditchfield, J. Singano, K. G. Harcourt-Brown, C. J. Nicholas, R. K. Olsson, N. J. Shackleton, and M. A. Hall (2001), Warm tropical sea surface temperatures in the Late Cretaceous and Eocene epochs, *Nature, 413*, 481–487, doi:10.1038/35097000.

Pike, C. R., A. P. Roberts, and K. L. Verosub (1999), Characterizing interactions in fine magnetic particle systems using first order reversal curves, *J. Appl. Phys., 85*, 6660–6667.

Rebesco, M., A. Camerlenghi, R. Geletti, and M. Canals (2006), Margin architecture reveals the transition to the modern Antarctic ice sheet (AIS) ca. 3 Ma, *Geology, 34*(4), 301–304, doi:10.1130/G22000.1.

Roberts, A. P., C. R. Pike, and K. L. Verosub (2000), First-order reversal curve diagrams: A new tool for characterizing the magnetic properties of natural samples, *J. Geophys. Res., 105*(B12), 28,461–28,475.

Smith, R. T., and J. B. Anderson (2010), Ice sheet evolution in James Ross Basin, Weddell Sea margin of the Antarctic Peninsula: The seismic stratigraphic record, *Geol. Soc. Am. Bull., 22*, 830–842.

Stoner, J. S., J. E. T. Channell, and C. Hillaire-Marcel (1996), The magnetic signature of rapidly deposited detrital layers from the deep Labrador Sea: Relationship to North Atlantic Heinrich layers, *Paleoceanography, 11*, 309–325.

Weeks, R., C. Laj, L. Endignoux, M. Fuller, A. P. Roberts, R. Manganne, E. Blanchard, and W. Goree (1993), Improvements in long-core measurement techniques: Applications in palaeomagnetism and palaeoceanography, *Geophys. J. Int., 114*, 651–662.

Wellner, J. S., J. B. Anderson, W. Ehrmann, F. M. Weaver, A. Kirshner, D. Livsey, and A. Simms (2011), History of an evolving ice sheet as recorded in SHALDRIL cores from the northwestern Weddell Sea, Antarctica, in *Tectonic, Climatic, and Cryospheric Evolution of the Antarctic Peninsula*, doi:10.1029/2010SP001047, this volume.

Wilson, G. S., A. P. Roberts, K. L. Verosub, F. Florindo, and L. Sagnotti (1998), Magnetobiostratigraphic chronology of the Eocene-Oligocene transition in the CIROS-1 core, Victoria Land margin, Antarctica: Implications for Antarctic glacial history, *Geol. Soc. Am. Bull., 110*, 35–47.

Yamazaki, T., and N. Ioka (1997), Cautionary note on magnetic grain-size estimation using the ratio of ARM to magnetic susceptibility, *Geophys. Res. Lett., 24*(7), 751–754.

Zachos, J. C., T. M. Quinn, and K. A. Salamy (1996), High-resolution (10^4 years) deep-sea foraminiferal stable isotope records of the Eocene-Oligocene climate transitions, *Paleoceanography, 11*(3), 251–266, doi:10.1029/96PA00571.

Zachos, J. C., M. Pagani, L. Sloan, E. Thomas, and K. Billups (2001), Trends, rhythms, and aberrations in global climate 65 Ma to Present, *Science, 292*, 686–693, doi:10.1126/science.1059412.

Zachos, J. C., T. Lee, and R. Kump (2005), Carbon cycle feedbacks and the initiation of Antarctic glaciation in the earliest Oligocene, *Global Planet. Change, 47*, 51–66.

L. Jovane, Instituto Oceanográfico da Universidade de São Paulo, Praça do Oceanográfico 191, São Paulo-SP, 05508-120, Brazil. (luigijovane@gmail.com)

K. L. Verosub, Department of Geology, University of California, Davis, Davis, CA 95616, USA.

History of an Evolving Ice Sheet as Recorded in SHALDRIL Cores From the Northwestern Weddell Sea, Antarctica

Julia S. Wellner,[1] John B. Anderson,[2] Werner Ehrmann,[3] Fred M. Weaver,[2] Alexandra Kirshner,[2] Daniel Livsey,[4] and Alexander R. Simms[4]

During 2006, the SHALDRIL program recovered cores of Eocene through Pliocene material at four locations in the northwestern Weddell Sea, each representing a key period in the evolution of the Antarctic Peninsula ice cap. The recovered cores are not continuous, yet they provide a record of climate change with samples from the late Eocene, late Oligocene, middle Miocene, and early Pliocene and represent the only series of samples recovered from the northwestern Weddell Sea and spanning the Cenozoic and the initial growth of the peninsula ice cap. Late Eocene sediments sampled in the James Ross Basin are typically characterized by very dark greenish-gray muddy fine sand with some preserved burrowing and are interpreted to represent a shallow water continental shelf setting. Rare dropstones, primarily of well-cemented sandstones and minor ice-rafted material consisting of angular grains with glacially influenced surface features record the onset of mountain glaciation, the earliest such evidence in the region. The remaining cores were collected on the Joinville Plateau to the north of the James Ross Basin. The late Oligocene sediments consist of dark gray sandy mud with some clay lenses and many burrows, likely representing a distal delta or shelf setting. This core contains only very few and small dropstones, and the individual grains show decreased angularity and fewer glacial surface features relative to late Eocene deposits. The middle Miocene strata are composed of pebbly gray diamicton, representing proximal glacimarine sediments. The lower Pliocene section also contains many ice-rafted pebbles but is dominated by sandy units rather than diamicton and is interpreted to represent a current-winnowed deposit, similar to the modern contour current-influenced sediments of the region.

[1]Department of Earth and Atmospheric Sciences, University of Houston, Houston, Texas, USA.
[2]Department of Earth Science, Rice University, Houston, Texas, USA.
[3]Institut für Geophysik und Geologie der Universität Leipzig, Leipzig, Germany.
[4]Department of Earth Science, University of California, Santa Barbara, California, USA.

Tectonic, Climatic, and Cryospheric Evolution of the Antarctic Peninsula
Special Publication 063
Copyright 2011 by the American Geophysical Union.
10.1029/2010SP001047

1. INTRODUCTION

An understanding of Antarctica's Cenozoic climate history has been limited by a paucity of both outcrop and sediment cores that sample strata representing this time interval. This gap in our detailed knowledge includes the time period when ice sheets were beginning to expand on the Antarctic continent. The history of this expansion is not only relevant to understanding climatic controls on ice sheets and their dynamics but is also germane to understanding the history of eustatic sea level changes and the resulting stratigraphic architecture [e.g., *Peters et al.*, 2010].

Five new sediment cores were obtained as part of the SHALDRIL program from the James Ross Basin and the Joinville Plateau, both located in the northwestern Weddell Sea (Figure 1). The cores sampled discrete periods of time in the late Eocene, late Oligocene, middle Miocene, and early Pliocene [*Bohaty et al.*, this volume]; one unconformity was recorded, separating the middle Miocene from the overlying early Pliocene. The cores record a transition from minor and/ or distal glaciation to more severe glacial conditions, culminating in the advance of the Antarctic Peninsula Ice Cap across the northernmost peninsula and adjacent offshore areas. The purpose of this chapter is to describe the sedimentology and clay mineralogy of the cores and document the climatic changes they record in comparison to onshore studies [e.g., *Smellie et al.*, 2008, 2009; *Ivany et al.*, 2008; *Marenssi et al.*, 2010] and offshore seismic studies [e.g., *Smith and Anderson*, 2010, this volume].

2. BACKGROUND

The James Ross Basin location, offshore of Seymour Island, was chosen for study because of the series of dipping strata, one of the most complete Neogene sections on the Antarctic margin. The onshore records of both James Ross and Seymour islands are quite well studied, at least by Antarctic standards. The Eocene La Meseta Formation is the youngest preglacial succession on the island [*Marenssi et al.*, 2002]. La Meseta deposition was followed by a long hiatus and then deposition of Plio-Pleistocene glacial units [*Gazdzicki et al.*, 2004].

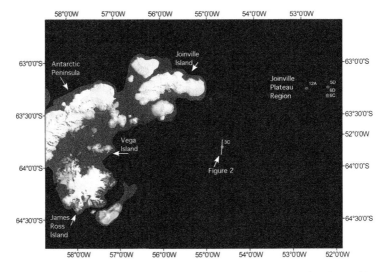

Figure 1. Map of northern tip of Antarctic Peninsula and Joinville Island with locations of SHALDRIL core sites marked. Location of seismic line in Figure 2 is also shown. Onshore image from http://lima.usgs.gov/.

Recently, *Ivany et al.* [2008] have used isotopic records from late Eocene bivalves from Seymour Island to indicate cooler temperatures in the late middle and late Eocene that may have been associated with a short-lived glacial advance. In addition, *Ivany et al.* [2006] have interpreted an outcrop on Seymour Island as glacial marine deposits or possibly till from the earliest Oligocene, but this outcrop may represent local mountain glaciers. *Marenssi et al.* [2010] questioned the possibility of Oligocene till on Seymour Island and reported late Miocene diamictites on Seymour Island, which they believe record the first in a series of glacial/interglacial cycles since that time.

A comprehensive sedimentological and geochronological analysis of volcanic and volcaniclastic deposits on James Ross and Vega Islands has led to a good record of glaciation in the northern peninsula region since the late Miocene [*Smellie et al.*, 2006, 2008; *Hambrey et al.*, 2008; *Nelson et al.*, 2009]. These works document repeated glacial/interglacial cycles prior to the late Miocene [~6.2 Ma; *Hambrey et al.*, 2008]. Their results call for a thin ice cover (200–350 m) interspersed with fewer periods of thick ice (600–750 m) as well as with interglacial periods [*Smellie et al.*, 2008]. Recent work on stable isotopic records of bivalves from the early Pliocene Cockburn Island Formation (~4.7 Ma) suggest that these marine organisms grew in sea ice-free water during an interglacial [*Williams et al.*, 2010], thus suggesting that the Pliocene must have included warm periods similar or perhaps warmer than present. However, *Johnson et al.* [2009] have suggested that the volcanic rocks on James Ross Island could not have been uncovered for more than 15,000 years out of the last 4.69 million years, based on cosmogenic dating of the volcanic surfaces. This would imply that ice cover persisted even during warm periods. From the offshore record, it is clear that the ice has been substantially farther advanced onto the shelf than its current position, although the timing of these grounding events is poorly constrained [e.g., *Anderson et al.*, 2002; *Smith and Anderson*, 2010].

The offshore portion of James Ross Basin was first studied by *Anderson et al.* [1992] and later by *Sloan et al.* [1995]. A recent paper by *Smith and Anderson* [2010] summarizes the stratigraphic history of the basin through analysis of multiple seismic data sets and an age model that is based on similarities in stratigraphic architecture to other parts of the margin where chronostratigraphic information exists. They argue that ice sheet development in the region proceeded gradually from south to north, with the first ice sheet grounding event on the continental shelf in the southern part of the basin having occurred in the late middle Miocene or early late Miocene and in the northern part of the basin in the early Pliocene.

Sea ice cover and rapidly drifting icebergs prevented drilling in the James Ross Basin to a few sites, only one of which sampled below the thick (~20 m) till that covers older strata. The remaining sites, which sampled Oligocene through Pliocene strata, were drilled on the Joinville Plateau area to the north, at the tip of the Antarctic Peninsula (Figure 1).

Since the middle Miocene, sedimentation in the region has been influenced by strong contour currents associated with the cyclonic circulation of the Weddell Gyre [*Maldonado et al.*, 2005]. Latest Pleistocene through Holocene contourite drifts occur along the eastern slope of the Weddell Sea [*Camerlenghi et al.*, 2001; *Howe et al.*, 2004], including the southern Joinville Plateau [*Smith et al.*, 2010]. The apparent intensification of contour currents has been tied to cooling and an increase in thermohaline currents in the area [*Michels et al.*, 2001, 2002]. Seismic stratigraphy off of the Joinville Plateau is characterized by a wedge of sediment that spans the late Oligocene through the early Pliocene with relatively constant sedimentation rates, which is consistent with contour current-influenced sedimentation [*Smith and Anderson*, this volume].

The currents of the Weddell Gyre have produced a winnowed surface at the seafloor and the sedimentary record is interpreted as stacked contour deposits [*Maldonado et al.*, 2005]. Seismic stratigraphy off of the Joinville Plateau indicates an uninterrupted wedge of sediment that spans the late Oligocene through the early Pliocene [*Smith and Anderson*, this volume]. The seismic data indicate relatively constant accumulations of sediment over time.

3. METHODS

The cores used in this analysis were collected in 2006 under the auspices of the SHALDRIL program. SHALDRIL was conceived in the 1990s as a way to obtain drill cores from the Antarctic continental margin by putting a drill rig onto an icebreaker, as a means to sample strata exposed close to the seafloor but in areas of regular sea ice cover. The first SHALDRIL test cruise took place in 2005 and was followed by a full cruise in the austral fall of 2006 (research vessel icebreaker *Nathaniel B. Palmer* cruise NBP0602A) during which the cores for this study were collected. Summaries of the drilling operations are provided by the SHALDRIL Shipboard Scientific Party (SHALDRIL 2005 Cruise Report, 2005, http://www.arf.fsu.edu/projects/documents/shaldril2005.pdf; SHALDRIL II 2006 NBP0602A Cruise Report, 2006, http://www.arf.fsu.edu/projects/documents/Shaldril_2_Report.pdf, hereinafter referred to as SHALDRIL Shipboard Scientific Party, cruise report, 2006), *Wellner et al.* [2005], and *Anderson et al.* [2006].

Each of the drill cores was run through a GeoTek multi-sensor core logger (MSCL) immediately after equilibration to room temperature after being brought on board the vessel. MSCL data collection included magnetic susceptibility, gamma density, and electrical resistivity (ER). *P* wave velocity data were also collected and used in correlating the longer cores to seismic records. Following the collection of the MSCL data, cores were split, and the visual lithology was described onboard including Munsell color code of the sediment color, smear slide description of the <250 μm fraction, and description of compositional, textural, and other sedimentologic characteristics. A handheld ER probe was then used as calibration of the MSCL ER measurements. Detailed descriptions of all onboard procedures are provided by SHALDRIL Shipboard Scientific Party (cruise report, 2006).

All sediment cores were transported in D-tubes to the Antarctic Research Facility at Florida State University where they are archived. Once there, the cores were X-rayed, and the radiographs were used to produce counts of pebbles. Pebbles that were visible at the surface of the core were washed and visually identified; other pebbles that were in the interior portion of the cores were left in place so as not to damage the core. Thus, while a quantitative assessment was made of the number of pebbles, a qualitative assessment was made of their lithology. Samples were also collected for grain size measurements and X-ray diffraction analysis of clay minerals. Grain size analysis was conducted on a Malvern laser particle size analyzer at Rice University. Samples were allowed to soak in water with sodium hexametaphosphate as a dispersant to break up clasts and flocculated clay particles before samples were added to the magnetic stirrer and measured. Duplicate measurements were made both of the same aliquot and of additional aliquots of the same sample.

X-ray diffraction sample processing and analyses followed standard procedures [e.g., *Ehrmann et al.*, 2005]. Bulk sediment samples were crushed, treated with 10% acetic acid and 5% H_2O_2 solution in order to remove carbonate and organic matter, respectively. The clay fraction (<2 μm) was separated in settling tubes. About 40 mg of clay was dispersed and mixed with an internal standard consisting of a 0.04% MoS_2 suspension. The samples were mounted as texturally oriented aggregates on aluminum tiles and solvated with ethylene-glycol vapor at 60°C. The samples were X-rayed (Rigaku Miniflex, CoKα radiation, 30 kV, 15 mA) in the range 3–40°2Θ with a scan speed of 0.02°2Θ/s. Additionally, the range 27.5–30.6°2Θ was measured with a step size of 0.01°2Θ in order to better resolve the (002) kaolinite peak and the (004) chlorite peak. Diffractograms were interpreted using the "MacDiff" software (freeware) (R. Petschick,

2001, http://www.geologie.uni-frankfurt.de/Staff/Homepages/Petschick/Classicsoftware.html). The main clay mineral groups illite, chlorite, kaolinite, and smectite are noted by their basal reflections. For semiquantitative evaluations of the mineral assemblages, empirically estimated weighting factors were used on the integrated peak areas of the individual clay mineral reflections [*Biscaye*, 1964, 1965; *Brindley and Brown*, 1980]. The crystallinity of smectite and illite is expressed as the integral breadth (IB). High values indicate poor crystallinities, whereas low values indicate good crystallinities. The composition of the illites can be estimated from the 5/10 Å peak area ratios [*Esquevin*, 1969] and the *d* values of the (001) illite peak.

Nineteen samples were analyzed for surface micromorphology and grain shape analysis. Quartz grains were separated from clays by soaking in a mixture of sodium hexametaphosphate overnight and gently wet-sieving through a 63 μm sieve. Samples were not sonicated to prevent altering of original textures. Samples were split, and grains were selected using a dissecting microscope. Quartz grains were mounted with carbon tape onto a scanning electron microscope stub for sample analysis following the procedures of *Mahaney et al.* [1988]. Samples were then sputter coated with approximately 20 nm of gold or carbon and examined using a FEI Quanta 400 high-resolution field emission scanning electron microscope at Rice University in high-vacuum mode. The composition of each grain was verified to be quartz using energy-dispersive X-ray spectroscopy. Ten grains in each were examined in detail for surface morphology (see *Kirshner and Anderson* [this volume] for details). Features were identified based on the criterion and examples from the work of *Mahaney* [2002]. Microtextures were recorded as not present, low abundance, medium abundance, and high abundance.

Twenty-one samples were analyzed for grain shape using Fourier shape analysis. Samples were first treated with hydrogen peroxide and hydrochloric acid to remove organics and carbonates, respectively. Quartz grains were then isolated from remaining feldspars and lithic fragments using lithium polytungstate, a heavy density fluid. Approximately two hundred quartz grains per sample were imaged and analyzed using a petrographic microscope and free imaging software for statistical determination of grain shape.

4. LITHOFACIES DESCRIPTIONS

4.1. James Ross Basin Site: Late Eocene

Site NBP0602A-3C was drilled offshore James Ross Island (Figure 1) and targeted a surface where the stratigraphic succession onlaps onto acoustic basement, and the dipping stratigraphic units are exposed relatively close to the seafloor (Figure 2). The total depth of the hole is 20 m below seafloor (mbsf) but with no recovery beyond 17 m subbottom depth. Onboard biostratigraphy indicated that units of two different ages were sampled with the contact just below 3 m. An upper unit of Holocene-Pleistocene till and glaciomarine material overlies late Eocene sediments (SHALDRIL Scientific Party, cruise report, 2006). Sr dating on shells from the matrix sediment, not from clasts of older material, further constrained the age to approximately 35.9 Ma or late Eocene [*Bohaty et al.*, this volume].

The dominant facies between 3 and 17 m, the late Eocene section, is poorly sorted, muddy to silty very fine sand with few subangular to subrounded pebbles (Figure 3). One large cobble (>15 cm) was recovered at 12.3 mbsf. This large cobble and the majority of the pebbles are calcite-cemented sandstone. Many of the sandstone clasts contain bivalve fragments. The sediment from 3 to 7.5 mbsf is greenish-black and from 7.5 to 17 mbsf is very dark gray to very dark greenish-gray. There are numerous gastropods and bivalves distributed throughout the core and minimal bioturbation. Smear slide analysis of the <250 μm fraction indicated variability of clay

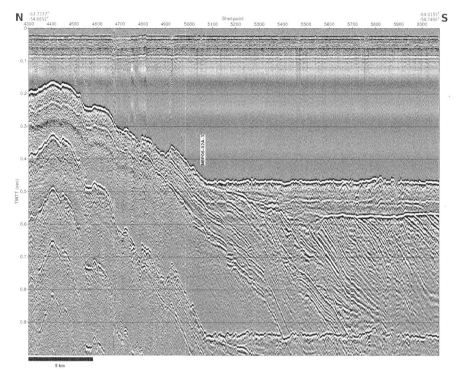

Figure 2. Seismic line NBP0602A-02 showing the surface where the condensed stratigraphic succession onlaps acoustic basement and the site of drill core NBP0602A-3C. The hole targeted the oldest stratigraphic unit where it came close to the surface, under thin Pleistocene glacial sedimentary cover. See Figure 1 for location.

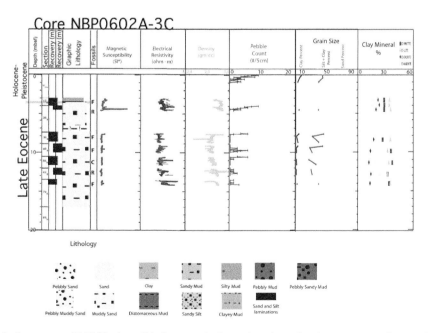

Figure 3. Summary of MSCL data, lithology, and clay mineralogy for the Eocene sediments in Hole NBP0602A-3C. Core locations are given in Figure 1.

content, ranging from 15% to 50% by volume with lowest concentration near the middle of the core. Quartz typically makes up 20%–30% of the sediment, while feldspars make up 5%–20%. Calcite varies from 0% to 40%, with a high in the area around 10.5 mbsf where clay minerals are at a low. Heavy minerals, including hornblende, glauconite, mica, and volcanic glass, make up 1%–2% of the sediment. Diatoms were found throughout the hole, as were sponge spicules that are typically pyritized. Silicoflagellates and radiolarians are distributed sporadically.

The late Eocene deposits recovered with Hole NBP0602A-3C are characterized by 26%–37% smectite, 32%–36% illite, 11%–23% chlorite, and 15%–23% kaolinite (Figure 3). The smectite is poorly crystalline (IB approximately 2.0 Δ°2Θ), while the illite is mainly well crystalline (IB 0.45–0.64 Δ°2Θ). The illite 5/10 Å ratio of 0.35–0.50 and the d values of 10.01–10.03 Å document the dominance of muscovite-like illites over biotite-like illites. The sediment clearly differs from that of the other drill sites to the west by its higher kaolinite and lower illite contents. In general, kaolinite and smectite decrease in relative concentration upcore, while chlorite increases.

Sand grains from the late Eocene sediments are highly angular with sharp edges and show little to no chemical weathering. The grains have surface textures that are compatible with glacial influence including striations, parallel fractures, and breakage blocks, which is in contrast to samples from upper part of the onshore La Meseta Formation, Seymour Island, which lacks evidence for glaciation [*Kirshner and Anderson*, this volume].

The MSCL data show that fine sand has densities from 1.7 to 2.5 g cm^{-3} with a slight but noticeable increase downhole. These densities are noticeably lower than the 2.3–2.7 g cm^{-3} in the younger unit above.

4.2. Joinville Plateau Sites

4.2.1. Oligocene. Hole NBP0602A-12A from the Joinville Plateau targeted the oldest acoustically laminated deposits that onlap strata that are nearly conformable with the basement below (see *Smith and Anderson* [this volume] for seismic data). The hole reached a total depth of 7.2 mbsf. Biostratigraphy indicates that the sediments of two different ages were sampled. Quaternary diatoms extend to at least 13 cm below the seafloor. Oligocene taxa were found from 15 to 7.2 m below seafloor, and the combined age ranges indicate an age for this section between 28.6 and 24.0 Ma [*Bohaty et al.*, this volume].

Hole NBP0602-12A sampled sediment varying between sandy mud and muddy sand and was divided into six lithologic units (Figure 4). Lithologic unit I extends from the seafloor to 14 cm (the Quaternary section) and consists of soft sandy mud with a few coarse pebbles and no macrofossils.

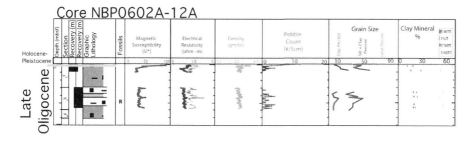

Figure 4. Summary of MSCL data, lithology, and clay mineralogy for Hole NBP0602A-12A, the Oligocene section. Core location is given in Figure 1, and lithologic symbols are the same as in Figure 3.

Lithologic units II through VI are all in the Oligocene section. Unit II extends from 0.14 to 2.90 mbsf and consists of firm very dark gray sandy mud that grades downward into firm very dark greenish-gray sandy mud (Figure 4). Lithologic unit III extends from 2.9 to 3.76 mbsf and consists of very dark gray muddy fine to very fine sand with no macrofossils. Diatoms compose 5% of the unit with sponge spicules and calcareous nannofossils in trace amounts. Lithologic unit IV extends from 3.76 to 4.20 mbsf and includes a layer of black muddy sand (3.92–4.00 mbsf) in between black sandy mud with no visible macrofossils. The sand throughout this unit is fine, and there are small lenses of clay throughout. Unit V extends from 4.20 to 5.35 mbsf and is composed of black muddy fine sand with small lenses of clay and several shell fragments. Lithologic unit VI extends from 5.35 to 7.20 mbsf and consists of sandy mud with no macrofossils and a couple of very small, subangular pebbles and one larger dropstone. Burrows are present throughout the late Oligocene section.

All samples from NBP0602A-12A, except the uppermost ones, show a relatively constant clay mineral composition with 23% smectite, 54% illite, 18% chlorite, and 5% kaolinite (Figure 4). Chlorite concentrations are about the same, and smectite concentrations are similar as in the upper part of NBP0602A-3C from James Ross Basin. As in the late Eocene sediments of James Ross Basin, the smectite is poorly crystalline (IB approximately 2.0 $\Delta°2\Theta$), the illite is well to moderately crystalline (IB 0.55–0.63 $\Delta°2\Theta$); the illite 5/10 Å ratio is 0.35–0.55. The most significant differences to the late Eocene sediments from James Ross Basin are the much lower kaolinite and the much higher illite contents.

Sand grains from the Oligocene section are semiangular with rounded edges. Large breakage blocks are the most common surface features on individual sand grains and are accompanied by some striations and fractures [*Kirshner and Anderson*, this volume]. Some silica precipitate is present. These grains show distinctly less angularity than the Eocene grains implying either less glacial influence and/or greater influence of marine currents.

Physical property measurements from this core indicate that the surface unit has a higher density (3.0 $g\,cm^{-3}$) than the muddy sand units below (~2.5 $g\,cm^{-3}$). Each layer of sandy mud is apparent on the density log by a spike in density values (Figure 4). Overall, each of the measured physical properties in this core is similar to those described below for the Miocene and Pliocene units nearby.

4.2.2. Miocene. Hole NBP0602A-5D was drilled in the Joinville Plateau area, in close proximity to the late Oligocene and Pliocene sites (Figure 1), and reached a maximum depth of 31.4 mbsf. Sediment was recovered between 8.0 mbsf and the base of the hole. Two unconformable contacts were sampled in the core (SHALDRIL Scientific Party, cruise report, 2006). The first, between 8.5 and 8.85 mbsf, separates surficial Quaternary deposits from early Pliocene deposits below (Figure 5). The second unconformity occurs at 18.80 mbsf and separates the Pliocene deposits from middle Miocene deposits below (SHALDRIL Scientific Party, cruise report, 2006). The Miocene deposits are described here, and the description of the Pliocene sediments is continued in the next section. *Bohaty et al.* [this volume] indicate that two different middle Miocene diatom zones were sampled at this site. The interval from 18.80 to 22.00 mbsf has an age of 12.1–11.8 Ma, and the unit below this, from 22.00 mbsf to the base of the cored interval, dates from 12.7 to 12.1 Ma.

The middle Miocene portion of Hole NBP0602A-5D is divided into three lithologic units, ranging from pebbly muddy sand to pebbly sandy diatomaceous mud. The upper unit (unit 1) extends from the top of the middle Miocene section to 22.18 mbsf. Lithologically, it is the same facies as the overlying early Pliocene deposits, so the unconformity that was identified by onboard biostratigraphy is not marked by a facies change. Unit 1 is characterized by very firm to stiff black

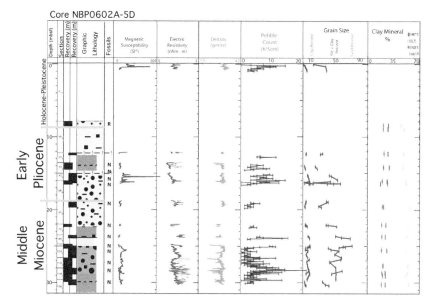

Figure 5. Summary of MSCL data, lithology, and clay mineralogy for Hole NBP0602A-5D, the Miocene section. Core location is given in Figure 1, and lithologic symbols are the same as in Figure 3.

to very dark greenish-gray pebbly muddy sand (Figure 5). The sand is dominantly fine and poorly sorted. Pebbles and sand grains are subangular to subrounded. There are no visible macrofossils in this interval, but it is organic rich, and fresh cut core yielded a hydrogen sulfide odor.

Unit II extends from 22.18 to 25.07 mbsf and is markedly finer than the unit above with very few sand grains, with the exception of lens of moderately sorted fine sand from 23.78 to 23.82 mbsf. The deposits are characterized as very dark gray, very firm silty mud with very few pebbles and no macrofossils.

Unit III extends from 25.07 to the base of the core at 31.4 mbsf and is similar in character to the upper Miocene unit. It is mostly greenish-black stiff sandy and silty mud with pebbles; the unit contains black mottles. Below 28.83 mbsf, the sediment is sandy diatomaceous mud. The sands are fine to very fine. There are no visible shells or bioturbation. This lowest unit is also characterized by thin white layers interpreted as gypsum. Physical property measurements show little downhole change, especially in density values that average 2.7 g cm^{-3} (Figure 5).

The middle Miocene sediments in the lower part of Hole NBP0602A-5D have almost exactly the same clay mineralogy as the late Oligocene sediments of core NBP0602A-12A: approximately 22% smectite, approximately 54% illite, approximately 18% chlorite, and approximately 7% kaolinite. Also, the crystallinities of smectite (IB approximately 1.8 $\Delta°2\Theta$), the crystallinity of illite (IB approximately 0.6 $\Delta°2\Theta$) and the illite 5/10 Å ratio (0.31–0.48) are almost identical.

Miocene sand grains are generally angular to subangular with rounded edges. Little chemical weathering occurs on grains. The surface features include gouges and fractures with very faint evidence for striations [see *Kirshner and Anderson*, this volume].

4.2.3. Pliocene. Early Pliocene sediments were recovered in the upper part of Hole NBP0602A-5D and in Holes NBP0602A-6C and NBP0602A-6D (Figure 1). Seismic data show that Site 6 is about 200 m up section stratigraphically from Site 5 [see *Smith and Anderson*, this volume]. As described above, Hole NBP0602A-5D includes Pliocene sediment, with an age range between 4.4 and 5.1 Ma, between two unconformities separating it from Miocene deposits below and Quaternary deposits above [*Bohaty et al.*, this volume]. NBP0602A-6C reached a total depth of 20.5 mbsf; the shallowest sediment recovered is from 4.0 mbsf (Figure 6). Two, or possibly three, pre-Quaternary biostratigraphic zones were sampled in this core. From 4.0 mbsf to 15.0 m, diatom analysis indicates an age between 4.4 and 3.7 Ma [*Bohaty et al.*, this volume]. From 15.0 to 17.0 m, the age of the unit is between 5.1 and 4.4 Ma [*Bohaty et al.*, this volume]. Below 17 m and continuing to the bottom of the core, the diatom zonation is not well constrained, and the age is either early Pliocene or late Miocene (SHALDRIL Scientific Party, cruise report, 2006). Hole NBP0602A-6D sampled Quaternary material above 1.0 mbsf and early Pliocene material (3.8– 4.4 Ma) below [*Bohaty et al.*, this volume], to a maximum depth of 10.0 mbsf. Because Site 6 is higher stratigraphically than Site 5, the material sampled within the 5.1–4.4 Ma diatom zone is most likely from the younger end of this age range and is continuous with the unit above, dating from 4.4 to 3.8 Ma. The age of the material in Site 5 cannot be constrained more tightly than 5.1– 4.4 Ma, but it has to be older than the material at Site 6.

The upper portion of the Pliocene section at Site NBP0602A-5D, from 8.0 to 12.15 mbsf (unit I), represents the oldest Pliocene sediments and is characterized by very dark greenish-gray medium to coarse sand and poorly sorted diamicton (Figure 5). The diamicton contains grain sizes ranging from clay through coarse sand with intermittent gravel layers and some large shell fragments. This grades downward into stiffer very dark gray muddy fine sand that lacks pebbles. From 12.15 to 15.0 mbsf (unit II), the deposits are firm sandy to silty diatomaceous mud that ranges in color from olive to black. From 15.0 to 18.8 mbsf (unit III), the site sampled stiff pebbly muddy sand varying in color from black to greenish-black. The sand is mostly fine to medium size.

The basal portion of Hole 6C (unit I), from 20.5 to 10.0 mbsf is the next youngest section after Site 5. This unit contains multiple diatom zones [*Bohaty et al.*, this volume] but is a single lithological facies characterized by firm dark gray to very dark greenish-gray pebbly muddy sand (Figure 6). The sand is dominantly fine and is rounded to subrounded. There is no indication of bioturbation or macrofossils. The upper portion of Hole 6C (unit II) extends from 4.0 to at least

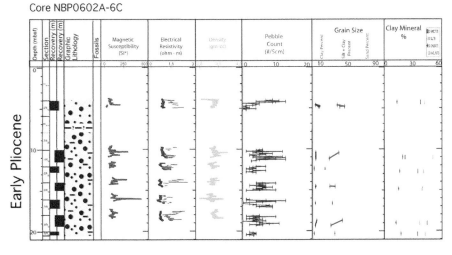

Figure 6. Summary of MSCL data, lithology, and clay mineralogy for Hole NBP0602A-6C, the Pliocene section. Core location is given in Figure 1, and lithologic symbols are the same as in Figure 3.

5.0 mbsf and potentially as deep as 10.0 mbsf (there is a gap in the core from 5.0 to 10.0), represents the youngest Pliocene material and consists of dark greenish-gray pebbly sand but differs from the unit below in that it also contains a few clay laminations.

The Pliocene material in Hole NBP0602A-6D occurs between 5.2 and 10.0 mbsf (unit II) and is similar to the units described from Hole 6C. The sediment consists primarily of dark greenish-gray to dark olive gray muddy sand to pebbly muddy sand. The sand is moderately sorted, fine to medium size, and there is no evidence for bioturbation or macrofossils. The pebbles contained in this unit have a wide range of lithologies.

Mean concentrations in the Pliocene section in the upper portion of Hole NBP0602A-5D are smectite 24%, illite 50%, chlorite 20%, and kaolinite 6%. The smectite is poorly crystalline (IB approximately 2.0 $\Delta°2\Theta$), the illite is well crystalline (IB 0.52-0.59 $\Delta°2\Theta$); the illite 5/10 Å ratio is 0.35–0.46. This clay mineral assemblage differs only marginally from that of the middle Miocene sediments of the same core by slightly enhanced smectite and chlorite concentrations and slightly lower illite concentrations. These values are similar to those from the younger Pliocene section from Hole NBP0602A-6C; however, the clay mineral percentages from Hole 6C show a much greater variability than the older Pliocene or Miocene sections.

The sand grains from the Pliocene section are rounded to subrounded, usually with rounded edges. Individual sand grains display some glacial surface features [see *Kirshner and Anderson*, this volume].

The MSCL logs for Hole NBP0602A-6C are dominated by spikes in MS (Figure 6). High values in MS can be correlated to concentrations in pebbles, particularly to the basaltic pebbles, which typically contain a high percentage of magnetic minerals. Similar to Hole NBP0602A-5D, the density log shows little downhole increase in density values, with an average density of around 2.8 g cm^{-3}.

5. DISCUSSION

The interpretation of the depositional history and climatic evolution recorded in the SHAL-DRIL cores is hampered by the fact that the sedimentary sequence under investigation is not continuous and, instead, is represented by a series of short cores, each of which contains gaps. Thus, the cores allow insight only into a few isolated stratigraphic intervals. Furthermore, the cores have been taken from different locations, both the James Ross Basin and the Joinville Plateau. Therefore, we have to reckon with different source areas from site to site and from time interval to time interval. However, considering that there are no other samples of these ages from this region, the SHALDRIL cores do provide important information about the onset of glacial conditions on the Antarctic Peninsula to compare to the glacial history elsewhere on the continent as well as climate changes recorded by deep-sea records. The sedimentary facies of the SHALDRIL cores, and the changing environments they record, are summarized in Figure 7.

The late Eocene is known as a time of widespread cooling and a major step toward glacial conditions in the Antarctic [e.g., *Lear et al.*, 2008] but the oldest well-constrained evidence for dates from the Oligocene in the northern Antarctic Peninsula [*Birkenmajer et al.*, 2005; *Ivany et al.*, 2006] and also from the South Shetland Islands [*Troedson and Smellie*, 2002]. Core NBP0602A-3C, in the James Ross Basin, dates from the late Eocene (see *Bohaty et al.* [this volume] for full details of the age model) and, based on seismic stratigraphic projections from the marine section to the onshore section, is believed to lie above the Eocene La Meseta Formation on nearby Seymour Island. Dates from the onshore La Meseta Formation, which has been notoriously difficult to date, have been given as 34.2 [*Dingle et al.*, 1998] and as no older than late

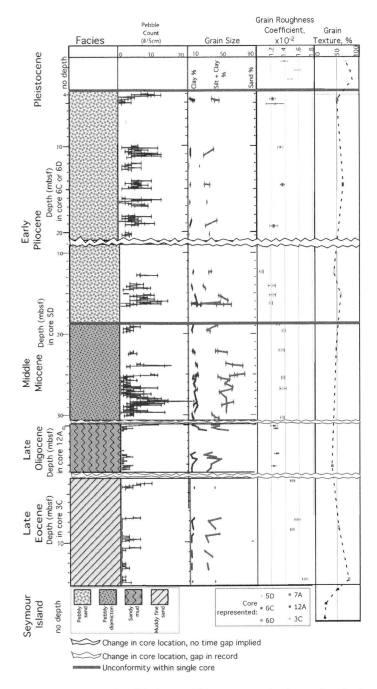

Figure 7. Summary diagram comparing lithology, pebble counts, grain size, and grain shape analysis for each core, from the Eocene through the Pliocene. Because this is a composite log, the *y* axis shows relative depth from each core; age overlaps have been cut and the Pleistocene-Modern overburden sampled in each core is not included except for comparison in grain roughness and textures. Grain roughness coefficients are calculated for harmonics 11–20. Seymour Island samples from the preglacial onshore strata are included for comparison in grain textures, also; see *Kirshner et al.* [this volume] for discussion.

Eocene [*Dutton et al.*, 2002]. Our age for this core of late Eocene and possibly 35.9 Ma [*Bohaty et al.*, this volume] suggests the possibility that the offshore unit is not truly above the onshore La Meseta but possibly coeval or slightly older even.

The Eocene sediments recovered in core NBP0602A-3C are typically characterized by very dark greenish-gray muddy fine sand with some burrows and are interpreted to represent a shallow water continental shelf setting [e.g., *Fielding et al.*, 2000], similar to the La Meseta section onshore Seymour Island [e.g., *Ivany et al.*, 2008]. Seismic stratigraphic work has shown that these deposits occur in isolated lobes, interpreted as deltas, that fringe eastern James Ross Basin [*Smith and Anderson*, 2010]. Rare dropstones, primarily well-cemented sandstones, represent ice-rafted clasts, as this is the only possible source of pebbles in this environment of deposition. Thus, while limited, the dropstones found in this core represent transport by ice was happening by the late Eocene, earlier than previously indicated for this region. Since most pebbles are of the same lithology, they are interpreted to represent a limited source area, possibly mountain glaciation. This is supported by the presence of glacial surface textures on sand grains [*Kirshner and Anderson*, this volume]. The location of the mountain glaciation is not clear as the exact provenance of the pebbles is unknown. The source of the ice-rafted sediments may be from islands close to the James Ross Basin or, our preferred interpretation, the ice may have come from ice on the Antarctic Peninsula mainland farther to the south. Terrestrial palynomorphs record southern beach-dominated and conifer forest in the Antarctic Peninsula at this time interval and indicate colder conditions than those recorded in the Eocene La Meseta Formation on Seymour Island [*Warny and Askin*, this volume(a)].

The interpretation of the clay fractions in this study is based on the observation that the clay mineral assemblages of sediments in the Antarctic region generally consist of illite, chlorite, smectite, and minor kaolinite. These minerals are mostly of detrital origin, and their composition is mainly controlled by the composition of the source rocks and by the type of weathering, i.e., chemical weathering under a humid and relatively warm climate versus physical weathering under a dry and cool climate.

Illite is the typical clay mineral of high latitudes and indicates the predominance of physical weathering on land. Although illite tends to be derived from acidic crystalline rocks, sediments such as sandstones and siltstones may also be possible source rocks. Illite cannot form in situ in the marine environment [*Biscaye*, 1965; *Griffin et al.*, 1968; *Windom*, 1976]. The late Eocene sediments from Hole NBP0602A-3C clearly differ from the younger sediments at other SHALDRIL sites by having higher kaolinite and lower illite contents, compatible with representing the preglacial stratigraphy. The Eocene clays of the Maud Rise and the South Orkney Plateau, to the north of the northwestern Weddell Sea, are dominated by smectite, which also indicates the continuation of chemical weathering processes in a relatively warm environment [*Barker et al.*, 1988].

Chlorite is a detrital clay mineral, like illite, and is common in high latitudinal settings. It is characteristic of physical weathering of low grade, chlorite-bearing metamorphic and basic rocks. The upcore increase in chlorite content (from 10% to 20%) within the Eocene section may indicate increasing physical weathering, since chlorite is destroyed by chemical weathering in warm and humid regions and does not survive transport to the oceans in significant amounts [*Petschick et al.*, 1996]. A similar increase in the chlorite concentration in late Eocene sediments on Maud Rise and southern Kerguelen Plateau was also interpreted to have been caused by intensified physical weathering [*Ehrmann and Mackensen*, 1992]. However, we cannot rule out that a change in the source area caused the increased chlorite concentrations at Site NBP0602A-3.

Smectite in marine sediments can either be of detrital or authigenic origin. Authigenic smectites originate from volcanism, hydrothermal activity, and/or diagenetic processes [e.g., *Chamley*,

1989; *Hillier*, 1995]. Detrital smectites, in contrast, are normally the result of chemical weathering under warm and humid climatic conditions in areas with very slowly moving water and contrasting wet and dry seasons [*Chamley*, 1989; *Robert and Chamley*, 1991]. However, high smectite concentrations in a cold climate with physical weathering commonly indicate a source area that is characterized by volcanic rocks [*Ehrmann et al.*, 1992; *Ehrmann*, 1998]. The poor smectite crystallinity of around 2.0 Δ°2Θ suggests that the smectites at Site NBP0602A-3 are derived from physical weathering. Similar crystallinities had been documented in the CIROS-1 and the Cape Roberts drill cores in McMurdo Sound, Ross Sea, in intervals dominated by physical weathering. In contrast, intervals influenced by chemical weathering and intervals containing common authigenic smectites had better smectite crystallinities [*Ehrmann*, 1998; *Ehrmann et al.*, 2005]. Appropriate volcanic source rocks are widespread in the Antarctic Peninsula region, although no volcanic rocks were found among the few pebbles contained in this core. The smectite concentrations decrease from approximately 35% to approximately 18% within the late Eocene section of Hole NBP0602A-3C. This decrease in the smectite content is likely due to a shift in provenance.

Kaolinite in marine sediments normally is a product of chemical weathering on the neighboring continents. High concentrations are restricted to temperate to tropical source regions with high, nonseasonal rainfall, where long continued and intense hydrolysis of granitic source rocks occurs in well-drained areas [*Chamley*, 1989; *Robert and Chamley*, 1991]. Thus, the high kaolinite contents in core NBP0602A-3C theoretically could be the result of humid conditions and hydrolysis on the Antarctic continent. The relatively high chlorite contents, however, contradict hydrolysis. Furthermore, the presence of some dropstones indicates at least mountain glaciation in the hinterland with some glaciers reaching the coast and, thus, also contradicts widespread chemical weathering conditions. Kaolinite cannot form under polar conditions. However, because kaolinite is a very resistant mineral, reworked kaolinite from older strata has frequently been found in polar environments [*Chamley*, 1989; *Hambrey et al.*, 1991; *Ehrmann et al.*, 1992, 2003; *Holmes*, 2000]. The most likely explanation for the high kaolinite content in the late Eocene sediments, therefore, is the erosion of sediments, which received their clay mineral signature under a warm and wet climate allowing kaolinite formation. Such conditions existed in early Paleogene and older time [e.g., *Robert and Maillot*, 1990; *Robert et al.*, 2002]. The slight upcore decrease in kaolinite concentrations may be due to a change in provenance. Our interpretation on the origin of the late Eocene kaolinite is confirmed by the fact that the uppermost investigated sample coming from the Quaternary interval of core NBP0602A-3C has an almost identical signature as the late Eocene sediments and that we can rule out chemical weathering for the Quaternary time interval (Figure 3).

Coeval sediments have been drilled in the Southern Ocean by the Ocean Drilling Program. The late Eocene kaolinite concentrations are only approximately 5% in ODP Site 696, in the South Orkney area [*Robert and Maillot*, 1990], ODP Site 690 on Maud Rise, and in ODP Sites 738 and 744 on the southern Kerguelen Plateau [*Ehrmann and Mackensen*, 1992]. In ODP Site 689 on Maud Rise, they are up to 10% [*Ehrmann and Mackensen*, 1992; *Robert et al.*, 2002]. The low kaolinite contents in these sites further supports the idea of a local kaolinite source for the sediments in James Ross Basin.

In the late Eocene part of the sedimentary sequence onshore Seymour Island, both kaolinite and smectite contents of approximately 40% each have been found, and both minerals show an upward-decreasing trend [*Dingle et al.*, 1998]. In contrast to the James Ross Basin, no chlorite and no ice-rafted detritus were detected in the onshore section [*Dingle et al.*, 1998]. However, the high kaolinite and smectite contents are accompanied by a chemical index of alteration that indicates cool temperatures.

The clay minerals of the late Eocene sediments of Site NBP0602A-3 are probably provided by physical weathering of a mixed source consisting of crystalline basement, old sedimentary rocks, and volcanic rocks. During the time represented by the core, the influence of the sedimentary rocks decreased, whereas the influence of the basement and volcanic rocks increased.

Each of the younger SHALDRIL sites was drilled off of the Joinville Plateau (Figure 1). Seismic data from this area show a highly laminated, seaward dipping succession lacking in any apparent unconformities with the exception of one that occurs near the top of the section [*Smith and Anderson*, this volume]. Site NBP0602A-12A sampled Oligocene strata between 28.6 and 24.0 Ma [*Bohaty et al.*, this volume]. The sediments consist of dark gray sandy mud with some clay lenses and many burrows throughout, likely representing a distal delta or lower shoreface setting. This core contains only one clear dropstone, and grain angularity and relatively small numbers of glacial surface textures on individual grains indicate minor glacial influence and/or a setting that is distal from the Antarctic Peninsula. While the Oligocene strata from the Joinville Plateau show relatively small amounts of glacial influence, elsewhere in the northern peninsula region, there is evidence for more significant glaciation at the same time period [e.g., *Troedson and Smellie*, 2002]. The Oligocene and younger SHALDRIL cores are from the Joinville Plateau and are thus much more distal from land, and area for ice sheet growth, than the Eocene core from the James Ross Basin. Thus, there may well have been significant expansion of ice cover from the Eocene to the Oligocene, but the sampling location for the Oligocene section is not proximal enough to have a strong record of glacial sediments. The palynomorph assemblage from Site 12 is dominated by recycled specimens and is of lower diversity than the late Eocene assemblage and, overall, indicates a much colder climate than that recorded at Site NBP0602A-3C [*Warny and Askin*, this volume(a)].

The difference in the clay mineral composition between the Eocene and Oligocene sites may be a result of different source areas rather than different weathering regimes. The illite- and chlorite-rich assemblage from Hole NBP0602A-12A is typical for physical weathering of a source dominated by crystalline basement [*Ehrmann*, 1998; *Ehrmann et al.*, 2005]. Volcanic rocks obviously are present in the hinterland, but less widespread; kaolinite-bearing sedimentary rocks are insignificant as source rocks.

The middle Miocene was a time of major climate change on the Antarctic continent with a permanent shift toward polar conditions [e.g., *Lewis et al.*, 2007]. Middle Miocene sediments were sampled at Hole NBP0602A-5D [*Bohaty et al.*, this volume]. The sediments are character-ized as a pebbly gray diamicton, representing proximal glacimarine sediments and represent a dramatic increase in glacial conditions relative to that recorded by the Oligocene strata at Site NBP0602A-12A. The pebbles contained within the diamicton are dominantly metasedimentary, likely indicating a single glacial source rather than iceberg rafting. Individual sand grains are more angular and have larger numbers of glacial surface textures relative to Oligocene deposits [*Kirshner and Anderson*, this volume]. The palynomorphs from this core are mainly recycled and record a limited taxa onshore with sea ice species offshore [*Warny and Askin*, this volume(b)].

The sedimentary facies indicates a major paleoenvironmental shift between the late Oligocene and the middle Miocene to a more glacial regime, yet this is not reflected in the clay mineral distribution. This may be explained by a persistent source area dominating the signal. The environmental shift apparently had no major effect on the weathering regime. Once physical weathering was established on land as the dominant process, further cooling did not cause a change in the weathering processes.

The lower Pliocene facies sampled at Site NBP0602A-6D, as well as a shorter section from Site NBP0602A-5D, like the Miocene section contains many ice-rafted pebbles but is dominated by sand units rather than diamicton and is interpreted to represent a current-winnowed glacimarine deposit, similar to the modern day sediment on the Joinville Plateau. The Pliocene sediments are composed of alternating zones of structureless fine and occasionally medium sand alternating with structureless muddy units that also contain fine sands, all of which are characteristic features of current deposits [*Stow and Lovell*, 1979]. The fact that the unit is formed through winnowing of glacimarine sediments results in higher concentrations of pebbles than contour deposits commonly have, but otherwise, the sedimentary features are quite diagnostic. This unit is lacking palynomorphs, suggesting no significant vegetation was present at this time [*Warny and Askin*, this volume b] and that polar conditions existed.

Figure 8 shows a comparison of grain size distribution for representative samples for each age interval sampled. The Eocene through Miocene samples have coarser, more poorly sorted grain size distributions, while the Pliocene samples show a prominent very fine sand mode. This is interpreted as indicating the onset of contour current-controlled deposition at this site between the middle Miocene and early Pliocene.

The MSCL logs from the Miocene and Pliocene section differ from the Oligocene and Eocene (Figures 3–6). The Miocene and particularly the Pliocene section have relatively constant densities that do not noticeably increase downcore, whereas the Eocene section shows a noticeable increase in densities downcore. If much of the finer grain material is removed from the sandier units, grain-to-grain contact may prevent compaction until greater burial, leaving winnowed deposits with relatively constant downcore densities. Despite the sandy nature of the Eocene section, the densities do increase downcore indicating that some compaction has taken

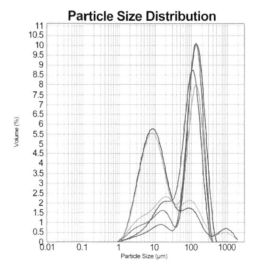

Figure 8. Grain size frequency for selected samples, from the <2 mm fraction. The red curve is from Hole 3C at 17.47 m from the Eocene strata; green curve is from Hole 12A at 3.85 m from the Oligocene strata; pink curve is from Hole 5D at 19.47 m in the Pliocene section; blue curve is from Hole 5D at 0.1 m in modern sediments. The fine tail (positive skewness) present in the Eocene and Oligocene section is not present in samples from the modern surface. The Pliocene samples (pink) show an intermediate level of sorting. The remaining two curves (army green from Hole 5D at 23.87 and burnt red from Hole 5D 24.87) are anomalous compared to all other samples observed in this study and have a coarse tail. They represent the upper part of the middle Miocene section in this core, just below the Pliocene-Miocene unconformity. These coursegrained deposits may be related to erosion across the unconformity in a broad section of possibly reworked sediments.

place; only the Miocene and Pliocene sections have density logs consistent with current-winnowed sands.

Despite the consistency between the Oligocene and Miocene clay mineralogies, the early Pliocene sediments of cores NBP0602A-6C and -6D have a distinctly different clay mineral signature with about 37% smectite, 42% illite, 15% chlorite, and 6% kaolinite. In contrast to core NBP0602A-5D, the concentrations show major fluctuations throughout cores NBP0602A-6C and -6D. Despite the difference in the concentrations of the individual clay minerals, the composition of illite and smectite is similar between the two sites. The smectite is poorly crystalline (IB approximately 2.0 $\Delta°2\Theta$), the illite is well crystalline (IB 0.46–0.51 $\Delta°2\Theta$), the illite 5/10 Å ratio is 0.46–0.51. The differences between the two early Pliocene sites may be explained by a greater contribution from a volcanic source area in NBP0602A-6C and NBP0602A-6D than at Site 5.

The proximal glacimarine facies in the Joinville Plateau middle Miocene section indicates that an ice sheet was grounded on Joinville Island and some parts of the Joinville Plateau at this time. This corresponds to the approximate timing of the onset of ice grounding events in the southern James Ross Basin based on seismic data [*Smith and Anderson*, 2010], although the expansion of the ice cap to produce an unconformity on the continental shelf did not occur until after the middle Miocene and prior to or during the early Pliocene, as recorded at sites NBP0602A-5D and NBP0602A-6C [*Smith and Anderson*, this volume]. This apparent discrepancy, that the sediments record proximal glacial-marine deposits prior to the formation of the major seismic unconformity, may indicate the expansion of the ice in a local area around Joinville Island prior to a more regional expansion and development of the unconformity. In addition, the ice may have grown at a relatively slow pace, being proximal to the core location for a long duration prior to further expansion. Alternatively, the middle Miocene sediments may represent distal rather than proximal glacial-marine deposition, with the coarse-grained nature of the deposit subsequently enhanced by contour currents. We prefer the first interpretation based on the overall nature of the sand grains and pebbles, and the particle size distribution within the unit.

6. CONCLUSIONS

The four SHALDRIL sites that recovered pre-Quaternary sediments allow the first sedimentological reconstruction of the transition from early glacial to polar conditions in the northwestern Weddell Sea, offshore of the Antarctic Peninsula. Late Eocene, approximately 35.9 Ma, sediments in the James Ross Basin indicate a shallow water shelf setting. Dropstones and sand grains with glacial surface textures, which are not present in late Eocene sediments on Seymour Island, indicate the onset of mountain glaciation somewhere in the region in the latest Eocene, although the location of the ice is not known. These samples provide the earliest evidence for the onset of glacial conditions in the region. Clay minerals from this site are probably derived through physical weathering of a mixed source consisting of crystalline basement, old sedimentary rocks, and volcanic rocks. Oligocene through Pliocene strata were recovered farther north and farther offshore from the Joinville Plateau. Oligocene strata from between 28.6 and 24.0 Ma consist of dark gray sandy mud with some clay lenses and many burrows throughout, likely representing a distal delta or outer shelf setting. This core contains only one dropstone greater than 4 mm and sand grains contain minor glacial surface textures, indicating only minor glacial influence. Clay mineralogies suggest a sediment source dominated by crystalline basement. Middle Miocene strata from between 12.7 and 11.8 Ma indicate the onset of glacial conditions proximal to the Joinville Plateau, but there is no offshore glacial unconformity of this age. The middle Miocene section is truncated by an unconformity that represents initial advance of the ice cap onto Joinville Plateau during the early Pliocene. Sediment above this unconformity have an age range between 5.1 and 3.8 Ma and are characterized by poor sorting, many dropstones, and glacial surface textures on sand grains. These sediments contain a sorted fine sand mode and are associated with a pebbly lag component that indicates deposition under the influence of marine currents. These Pliocene current-influenced deposits are similar to modern sediments of the area, which are known to be influenced by contour currents associated with the Weddell Gyre.

Acknowledgments. This work was funded by NSF-OPP-0125922 to J.B.A. Thanks are extended to the science parties and crews of two cruises, NBP0502 and NBP0602A. Sylvia Dorn is thanked for assistance in completing the X-ray analysis. This manuscript benefited from thorough reviews by two anonymous reviewers.

REFERENCES

Anderson, J. B., S. S. Shipp, and F. P. Siringan (1992), Preliminary seismic stratigraphy of the northwestern Weddell Sea continental shelf, in *Recent Progress in Antarctic Earth Science*, edited by Y. Yoshida et al., pp. 603–612, Terra Sci., Tokyo.

Anderson, J. B., S. S. Shipp, A. L. Lowe, J. S. Wellner, and A. Mosola (2002), The Antarctic ice sheet during the last glacial maximum and its subsequent retreat history: A review, *Quat. Sci. Rev.*, *21*, 49–70.

Anderson, J. B., J. S. Wellner, S. Bohaty, P. L. Manley, and S. W. Wise Jr. (2006), Antarctic Shallow Drilling project provides key core samples, *Eos Trans. AGU*, *87*(39), 402, doi:10.1029/2006EO390003.

Barker, P. F., J. P. Kennet, and Leg 113 Scientific Party (1988), Weddell Sea Palaeoceanography: Preliminary Results of ODP Leg 113, *Palaeogeogr. Palaeoclimatol. Palaeoecol.*, *67*, 75–102.

Birkenmajer, K., A. Gazdzicki, K. P. Krajewski, A. Przybycin, A. Solecki, A. Tatur, and H. I. Yoon (2005), First Cenozoic glaciers in West Antarctica, *Pol. Polar Res.*, *26*, 3–12.

Biscaye, P. E. (1964), Distinction between kaolinite and chlorite in recent sediments by X-ray diffraction, *Am. Mineral.*, *49*, 1281–1289.

Biscaye, P. E. (1965), Mineralogy and sedimentation of recent deep-sea clay in the Atlantic Ocean and adjacent seas and oceans, *Geol. Soc. Am. Bull.*, *76*, 803–832.

Bohaty, S. M., D. K. Kulhanek, S. W. Wise Jr., K. Jemison, S. Warny, and C. Sjunneskog (2011), Age assessment of Eocene–Pliocene drill cores recovered during the SHALDRIL II expedition, Antarctic Peninsula, in *Tectonic, Climatic, and Cryospheric Evolution of the Antarctic Peninsula*, doi:10.1029/2010SP001049, this volume.

Brindley, G. W., and G. Brown (Eds.) (1980), Crystal structures of clay minerals and their X-ray identification, , *Mineral. Soc. Monogr.*, *5*, 495 pp.

Camerlenghi, A., E. Domack, M. Rebesco, R. Gilbert, S. Ishman, A. Leventer, S. Brachfeld, and A. Drake (2001), Glacial morphology and post-glacial contourites in northern Prince Gustav Channel (NW Weddell Sea, Antarctica), *Mar. Geophys. Res.*, *22*, 417–443.

Chamley, H. (1989), *Clay Sedimentology*, 623 pp., Springer, Berlin.

Dingle, R. V., S. A. Marenssi, and M. Lavelle (1998), High latitude Eocene climate deterioration: Evidence from the northern Antarctic Peninsula, *J. South Am. Earth Sci.*, *11*, 571–579.

Dutton, A. L., K. C. Lohmann, and W. J. Zinsmeister (2002), Stable isotope and minor element proxies for Eocene climate of Seymour Island, Antarctica, *Paleoceanography*, *17*(2), 1016, doi:10.1029/2000PA000593.

Ehrmann, W. (1998), Implications of late Eocene to early Miocene clay mineral assemblages in McMurdo Sound (Ross Sea, Antarctica) on paleoclimate and ice dynamics, *Palaeogeogr. Palaeoclimatol. Palaeoecol.*, *139*, 213–231.

Ehrmann, W., and A. Mackensen (1992), Sedimentological evidence for the formation of an East Antarctic ice sheet in Eocene/Oligocene time, *Palaeogeogr. Palaeoclimatol. Palaeoecol.*, *93*, 85–112.

Ehrmann, W., M. Melles, G. Kuhn, and H. Grobe (1992), Significance of clay mineral assemblages in the Antarctic Ocean, *Mar. Geol.*, *107*, 249–273.

Ehrmann, W., J. Bloemendal, M. J. Hambrey, B. McKelvey, and J. Whitehead (2003), Variations in the composition of the clay fraction of the Cenozoic Pagodroma Group: Implications for determining provenance, *Sediment. Geol.*, *161*, 131–152.

Ehrmann, W., M. Setti, and L. Marinoni (2005), Clay minerals in Cenozoic sediments off Cape Roberts (McMurdo Sound, Antarctica) reveal palaeoclimatic history, *Palaeogeogr. Palaeoclimatol. Palaeoecol.*, *229*, 187–211.

Esquevin, J. (1969), Influence de la composition chimique des illites sur le cristallinité, *Bull. Cent. Rech. Pau*, *3*, 147–154.

Fielding, C. R., T. R. Naish, K. J. Woolfe, and M. A. Lavelle (2000), Facies analysis and sequence stratigraphy of CRP-2/2A, Victoria Land Basin, Antarctica, *Terra Antart.*, *7*, 323–338.

Gazdzicki, A., A. Tatur, U. Hara, and R. A. del Valle (2004), The Weddell Sea formation: Post-late Pliocene terrestrial glacial deposits on Seymour Island, Antarctic Peninsula, *Pol. Polar Res.*, *25*, 189–204.

Griffin, J. J., H. Windom, and E. D. Goldberg (1968), The distribution of clay minerals in the World Ocean, *Deep Sea Res. Oceanogr. Abstr.*, *15*, 433–459.

Hambrey, M. J., W. U. Ehrmann, and B. Larsen (1991), Cenozoic glacial record of the Prydz Bay continental shelf, East Antarctica, *Proc. Ocean Drill. Program Sci. Results*, *119*, 77–132.

Hambrey, M. J., J. L. Smellie, and A. E. Nelson (2008), Late Cenozoic glacier-volcano interaction on James Ross Island and adjacent areas, Antarctic Peninsula region, *Geol. Soc. Am. Bull.*, *120*, 709–731, doi:10.1130/B26242.1.

Hillier, S. (1995), Erosion, sedimentation and sedimentary origin of clays, in *Origin and Mineralogy of Clays*, edited by B. Velde, pp. 162–219, Springer, Berlin.

Holmes, M. A. (2000), Clay mineral composition of glacial erratics, McMurdo Sound, in *Paleobiology and Paleoenvironments of Eocene Rocks, McMurdo Sound, East Antarctica*, Antarct. Res. Ser., vol. 76, edited by J. D. Stilwell and R. M. Feldmann, pp. 63–71, AGU, Washington, D. C.

Howe, J. A., T. M. Shimmield, and R. Diaz (2004), Deep-water sedimentary environments of the north-western Weddell Sea and South Sandwich Islands, Antarctica, *Deep Sea Res., Part II*, *51*, 1489–1514.

Ivany, L. C., S. Van Simaeys, E. W. Domack, and S. D. Samson (2006), Evidence for an earliest Oligocene ice sheet on the Antarctic Peninsula, *Geology*, *34*, 377–380, doi:10.1130/G22383.1.

Ivany, L. C., K. C. Lohmann, F. Hasiuk, D. B. Blake, A. Glass, R. B. Aronson, and R. M. Moody (2008), Eocene climate record of a high southern latitude continental shelf: Seymour Island, Antarctica, *Geol. Soc. Am. Bull.*, *120*, 659–678, doi:10.1130/B26269.1.

Johnson, J. S., J. L. Smellie, A. E. Nelson, and F. M. Stuart (2009), History of the Antarctic Peninsula Ice Sheet since the early Pliocene—Evidence from cosmogenic dating of Pliocene lavas on James Ross Island, Antarctica, *Global Planet. Change*, *69*, 205–213.

Kirshner, A. E., and J. B. Anderson (2011), Cenozoic glacial history of the northern Antarctic Peninsula: A micromorphological investigation of quartz sand grains in *Tectonic, Climatic, and Cryospheric Evolution of the Antarctic Peninsula*, doi:10.1029/2010SP001046, this volume.

Lewis, A. R., D. R. Marchant, A. C. Ashworth, S. R. Hemming, and M. L. Machlus (2007), Major middle Miocene global climate change: Evidence from East Antarctica and the Transantarctic Mountains, *Geol. Soc. Am. Bull.*, *119*, 1449–1461, doi:10.1130/0016-7606(2007)119[1449:MMMGCC]2.0.CO;2.

Mahaney, W. C. (2002), *Atlas of Sand Grain Surface Textures and Applications*, 256 pp., Oxford Univ. Press, Oxford, U. K.

Mahaney, W. C., W. Vortisch, and P. Julig (1988), Relative differences between glacially-crushed quartz transported by mountain and continental ice: Some examples from North America and East Africa, *Am. J. Sci.*, *288*(10), 810–826.

Maldonado, A., et al. (2005), Miocene to recent contourite drifts development in the northern Weddell Sea (Antarctica), *Global Planet. Change*, *45*, 99–129, doi:10.1016/j.gloplacha.2004.09.013.

Marenssi, S. A., L. I. Net, and S. N. Santillana (2002), Provenance, environmental and paleogeographic controls on sandstone composition in an incised-valley system: The Eocene La Meseta Formation, Seymour Island, Antarctica, *Sediment. Geol.*, *150*, 301–321, doi:10.1016/S0037-0738(01)00201-9.

Marenssi, S. A., S. Casadio, and S. N. Santillana (2010), Record of late Miocene glacial deposits on Isla Marambio (Seymour Island), Antarctic Peninsula, *Antarct. Sci.*, *22*(2), 193–198.

Michels, K. H., J. Rogenhagen, and G. Kuhn (2001), Recognition of contour-current influence in mixed contourites-turbidite sequences of the western Weddell Sea, Antarctica, *Mar. Geophys. Res.*, *22*, 465–485.

Michels, K. H., G. Kuhn, C.-D. Hillenbrand, B. Diekmann, D. K. Fütterer, H. Grobe, and G. Uenzelmann-Neben (2002), The southern Weddell Sea: Combined contourites-turbidite sedimentation at the southeastern margin of the Weddell Gyre, in *Deep Water Contourite Systems: Modern Drifts and Ancient Series, Seismic and Sedimentary Characteristics*, edited by D. A. Stow et al., *Geol. Soc. Spec. Publ.*, *22*, 305–323.

Nelson, A. E., J. L. Smellie, M. J. Hambrey, M. Williams, M. Vautravers, U. Salzmann, J. M. McArthur, and M. Regelous (2009), Neogene glacigenic debris flows on James Ross Island, northern Antarctic Peninsula, and their implications for regional climate history, *Quat. Sci. Rev.*, *28*, 3138–3160.

Peters, S. E., A. E. Carlson, D. C. Kelly, and P. D. Gingerich (2010), Large-scale glaciation and deglaciation of Antarctica during the late Eocene, *Geology*, *38*, 723–726, doi:10.1130/G31068.1.

Petschick, R., G. Kuhn, and F. Gingele (1996), Clay mineral distribution in surface sediments of the South Atlantic: Sources, transport, and relation to oceanography, *Mar. Geol.*, *130*, 203–229.

Robert, C., and H. Chamley (1991), Development of early Eocene warm climates, as inferred from clay mineral variations in oceanic sediments, *Palaeogeogr. Palaeoclimatol. Palaeoecol.*, *89*, 315–331.

Robert, C., and H. Maillot (1990), Paleoenvironments in the Weddell Sea area and Antarctic climates, as deduced from clay mineral associations and geochemical data, ODP Leg 113, *Proc. Ocean Drill. Program Sci. Results*, *113*, 51–70.

Robert, C., L. Diester-Haass, and H. Chamley (2002), Late Eocene–Oligocene oceanographic development at the southern high latitudes, from terrigenous and biogenic particles: A comparison of Kerguelen Plateau and Maud Rise, ODP sites 744 and 689, *Mar. Geol.*, *191*, 37–54.

Sloan, B. J., L. A. Lawver, and J. B. Anderson (1995), Seismic stratigraphy of Palmer Basin, in *Geology and Seismic Stratigraphy of the Antarctic Margin*, *Antarct. Res. Ser.*, vol. 68, edited by A. K. Cooper et al., pp. 235–260, AGU, Washington, D. C.

Smellie, J. L., J. M. McArthur, W. C. McIntosh, and R. Esser (2006), Late Neogene interglacial events in the James Ross Island region, *Palaeogeogr. Palaeoclimatol. Palaeoecol.*, *242*, 168–187.

Smellie, J. L., J. S. Johnson, W. C. McIntosh, R. Esser, M. T. Gudmundsson, M. J. Hambrey, and B. van Wyk de Vries (2008), Six million years of glacial history recorded in the James Ross Island Volcanic Group, Antarctic Peninsula, *Palaeogeogr. Palaeoclimatol. Palaeoecol.*, *260*, 122–148, doi:10.1016/j.palaeo.2007.08.011.

Smellie, J. L., A. M. Haywood, D.-D. Hillenbrand, D. J. Lunt, and P. J. Valdes (2009), Nature of the Antarctic Peninsula Ice Sheet during the Pliocene: Geological evidence and modelling results compared, *Earth Sci. Rev.*, *94*, 79–94.

Smith, R. T., and J. B. Anderson (2010), Ice-sheet evolution in James Ross Basin, Weddell Sea margin of the Antarctic Peninsula: The seismic stratigraphic record, *Geol. Soc. Am. Bull.*, *122*, 830–842, doi:10.1130/B26486.1.

Smith, R. T., and J. B. Anderson (2011), Seismic stratigraphy of the Joinville Plateau: Implications for regional climate evolution, in *Tectonic, Climatic, and Cryospheric Evolution of the Antarctic Peninsula*, doi:10.1029/2010SP000980, this volume.

Smith, J. A., C.-D. Hillenbrand, C. J. Pudsey, C. S. Allen, and A. G. C. Graham (2010), The presence of polynyas in the Weddell Sea during the last glacial period with implications for the reconstruction of sea-ice limits and ice sheet history, *Earth Planet. Sci. Lett.*, *296*, 287–298, doi:10.1016/j.epsl.2010.05.008.

Stow, D. A. V., and J. P. B. Lovell (1979), Contourites: Their recognition in modern and ancient sediments, *Earth Sci. Rev.*, *14*, 251–291.

Troedson, A. L., and J. L. Smellie (2002), The Polonez Cove Formation of King George Island, Antarctica: Stratigraphy, facies and implications for mid-Cenozoic cryosphere development, *Sedimentology*, *49*, 277–301.

Warny, S., and R. Askin (2011a), Vegetation and organic-walled phytoplankton at the end of the Antarctic greenhouse world: Latest Eocene cooling events, in *Tectonic, Climatic, and Cryospheric Evolution of the Antarctic Peninsula*, doi:10.1029/2010SP000965, this volume.

Warny, S., and R. Askin (2011b), Last remnants of Cenozoic vegetation and organic-walled phytoplankton in the Antarctic Peninsula's icehouse world, in *Tectonic, Climatic, and Cryospheric Evolution of the Antarctic Peninsula*, doi:10.1029/2010SP000996, this volume.

Wellner, J. S., J. B. Anderson, and S. W. Wise (2005), The inaugural SHALDRIL expedition to the Weddell Sea, Antarctica, *Sci. Drill.*, *1*, 40–43.

Williams, M., et al. (2010), Sea ice extent and seasonality for the early Pliocene northern Weddell Sea, *Palaeogeogr. Palaeoclimatol. Palaeoecol.*, *292*, 306–318, doi:10.1016/j.palaeo.2010.04.003.

Windom, H. L. (1976), Lithogenous material in marine sediments, in *Chemical Oceanography*, vol. *5*, edited by J. P. Riley and R. Chester, pp. 103–135, Academic, New York.

J. B. Anderson, A. Kirshner, and F. M. Weaver, Department of Earth Science, Rice University, Houston, TX 77005, USA.

W. Ehrmann, Institut für Geophysik und Geologie der Universität Leipzig, D-04103, Leipzig, Germany.

D. Livsey and A. R. Simms, Department of Earth Science, University of California, Santa Barbara, Santa Barbara, CA 93106, USA.

J. S. Wellner, Department of Earth and Atmospheric Sciences, University of Houston, Houston, TX 77204, USA. (jwellner@uh.edu)

Cenozoic Glacial History of the Northern Antarctic Peninsula: A Micromorphological Investigation of Quartz Sand Grains

Alexandra E. Kirshner and John B. Anderson

Department of Earth Science, Rice University, Houston, Texas, USA

Glacial transport, owing to its high shear-stress regime, imparts a unique suite of microtextures on quartz grains. Here we examine surface textures of quartz sand grains from expedition SHALDRIL I and II and from an outcrop on Seymour Island, Antarctic Peninsula. The samples range in age from Eocene to Pleistocene and provide a record of the onset of glaciation regionally. The Eocene La Meseta Formation, Seymour Island, is void of any high-stress microtextures, supporting earlier interpretations that this unit is free of glacial influence. The inception of glacially derived high-stress microtextures and a very low occurrence of subparallel linear fractures on quartz grains begins in the late Eocene, marking the onset of alpine glaciation. Oligocene grains exhibit a continued presence of glacially derived microtextures, with a similar style to the late Eocene sediments. The morphological character changes in the middle Miocene. The middle Miocene microtextures are characteristic of transport from large ice sheets, displaying an increase in high-stress microtextures such as grooves, deep troughs, and crescentic gouges, an elevated degree of physical weathering, and an increased abundance of subparallel linear fractures. This is due to larger transit distances by ice rafting from the West Antarctic Ice Sheet. The Pliocene and Pleistocene samples contain abundant glacial microtextures, consistent with other evidence for the existence of the northern Antarctic Peninsula Ice Sheet at this time.

1. INTRODUCTION

It has long been established that quartz grains are recorders of environmental history [*Krinsley and Doornkamp*, 1973; *Bull*, 1981; *Marshall*, 1987; *Mahaney*, 2002], as their robust nature allows for the signature of transportation to be imprinted onto their surface. This has led to an extensive body of literature based on analysis of such features using scanning electron microscopy (SEM). Recognition of past episodes of glacial erosion and transport is possible by identifying a suite of microtextures that are unique to sustained high-stress conditions [*Krinsley and Margolis*, 1969; *Mahaney and Kalm*, 1995, 2000; *Mahaney*, 2002; *Sweet and Soreghan*, 2010].

Tectonic, Climatic, and Cryospheric Evolution of the Antarctic Peninsula
Special Publication 063
10.1029/2010SP001046

The scientific aim of SHALDRIL was to investigate the long-term climate record of the Antarctic Peninsula region [*Anderson et al.*, 2011]. A number of sites were drilled which span the late Eocene to Pleistocene. In this chapter, we report results from an investigation of the micromorphology of quartz grains from the SHALDRIL site sediments in conjunction with Eocene sediments from neighboring Seymour Island in order to document major stages in Cenozoic glacial history of the northern Antarctic Peninsula region.

2. GEOLOGIC SETTING

This study area is located between 63°S to 65°S and 58°W to 52°W and includes Seymour Island, the northern James Ross Basin, and the southern flank of the Joinville Plateau, south of Joinville Island (Figure 1). Samples from Seymour Island were taken from the Eocene La Meseta Formation, in particular, the youngest Submeseta unit. This sandy unit has been interpreted as preglacial, based on an absence of dropstones or other evidence for glaciation [*Marenssi et al.*, 2002]. The northern James Ross Basin sites include Pleistocene till that occurs just below the seafloor, in a region of spectacular glacial geomorphic features [*Heroy and Anderson*, 2005], and an uppermost Eocene muddy sand unit sampled at SHALDRIL Site 3C (Figure 1). The Joinville Plateau sites (12A, 7A, 5D, 6D, and 6C) sampled a thick sedimentary wedge along the southern flank of the plateau and recovered sediments that range in age from late Oligocene to Pleistocene. These sites occur in an area that is influenced by contour currents from the Weddell Gyre [*Maldonado et al.*, 2005; *Wellner et al.*, this volume].

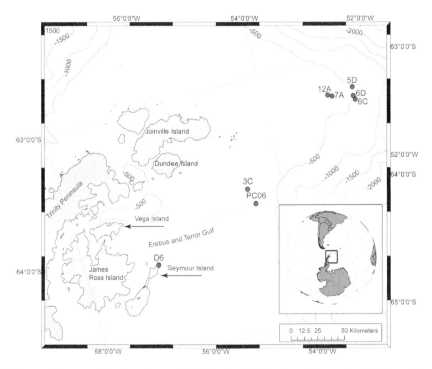

Figure 1. Geographic map of northern Antarctic Peninsula with locations of core sites. Bathymetric contours based on ETOPO2 [*Smith and Sandwell*, 1997] (see http://topex.ucsd.edu/marine_topo/mar_topo.html). Circles mark core sites (PC06, 3C, 12A, 7A, 5D, 6D, and 6C) and outcrop area (D6).

3. METHODS

A total of 10 sand grains per sample from 19 stratigraphic intervals were examined in detail for surface morphological characterization [*Stickley et al.*, 2009]. This includes 40 grains from the Holocene/Pleistocene, 20 grains from the Pleistocene, 20 grains from the early Pliocene, 20 grains from the middle Miocene, 20 grains from the late Oligocene, 30 grains from the late Eocene, and 30 grains from the Eocene. Rock samples from Seymour Island were provided by the United States Polar Rock Repository (D6-03, D6-05, D6-07). These samples were collected in 1986 and were noted to have been collected from pristine outcrops of unconsolidated sand. All other samples were acquired from drill core and piston cores.

Throughout the sample preparation, care was taken to preserve sample morphology and avoid generating any surface features, following the procedures of *Mahaney et al.* [1988]. Samples were prepared by first immersing approximately 0.5 g of bulk sediment into a mixture of sodium hexametaphosphate and reverse osmosis water for 24 h in order to deflocculate clays. Following disaggregation, the samples were gently wet-sieved through a 63 µm sieve. The sand-sized particles were then split, and quartz grains were selected using a dissecting microscope. The separated quartz grains were mounted onto SEM sample mounts with carbon tape. The mounted specimens were then sputtered-coated with approximately 20 nm of a conductive material (gold or carbon) and examined using a FEI Quanta 400 high-resolution field emission scanning electron microscope in high-vacuum mode. The composition of each grain was verified as pure SiO_2 with energy-dispersive X-ray spectroscopy.

Textural features were identified based on the criteria and examples from *Mahaney* [2002]. Microtextures were recorded as not-present (barren), low abundance (faint or scarce occurrence), medium abundance (present), and high abundance (exceptional occurrence). Fifteen different features were assessed in groups based on the process of microtexture formation [*Sweet and Soreghan*, 2010]. The groups include polygenetic features comprised of fracture faces, subparallel linear fractures, conchoidal fractures, arc-shaped steps, linear steps, sharp angular features; sustained high-stress features consisting of crescentic gouges, straight grooves, curved grooves, deep troughs, and mechanically upturned plates; percussion features of v-shaped percussion cracks and edge rounding; chemical dissolution/diagenesis features such as dissolution etching; and diagenetic physical weathering resulting in a weathered surface. Examples of some of the most diagnostic features are illustrated in Figure 2.

4. RESULTS

4.1. Eocene

4.1.1. Eocene La Meseta Formation, Seymour Island. The Eocene La Meseta Formation sites (D6-3m, D6-5m, and D6-7m, Figure 3) are characterized by a general low abundance of micro-textures. The most prevalent features are those associated with weathering and diagenesis, including weathered surfaces, dissolution etching, and edge rounding, which all occur in medium abundances. Over two thirds of the grains are rounded. Sustained high-impact features are present in low (crescentic gouges) to zero abundance (straight and curved grooves, deep troughs, and mechanically upturned plates) (Figure 4).

4.1.2. Late Eocene Offshore James Ross Basin. Late Eocene sand grains from SHALDRIL Site 3C (3C-3_8-9cm, 3C-3_114-115cm, and 3C-7_47-48cm, Figure 5) are fresh, with over half angular grains and a large majority of the grains displaying angular edge features. The most

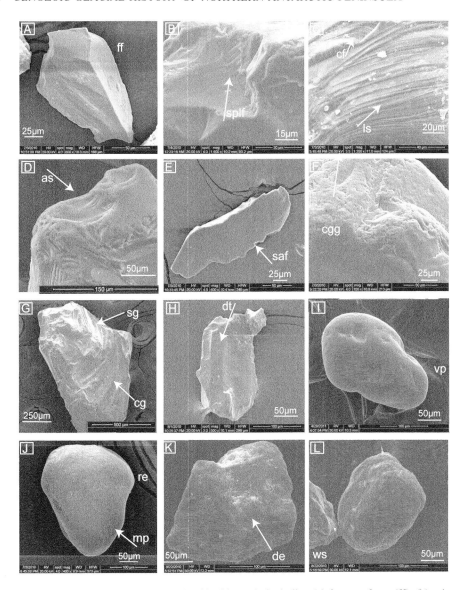

Figure 2. Examples of microtextures examined in this study including (a) fracture faces (ff), (b) subparallel linear fractures (splf), (c) (top left) conchoidal fractures (cf) and (right) linear steps (ls), (d) arc steps (as), (e) sharp angular features (saf), (f) crescentic gouges (cgg), (g) (top) straight grooves (sg) and (bottom) curved grooves (cg), (h) deep trough (dt), (i) v-shaped percussion pits (vp), (j) mechanically upturned plates (mp) and rounded edges (re), (k) dissolution etching (de), and (l) weathered surface (ws).

common microtexture is fracture faces, which occurs at medium frequency. The subsequent most abundant features include a medium to low abundance of linear steps and sharp angular features. The samples contain low abundances of the sustained high-impact features, including, in order of abundance, straight grooves, crescentic gouges, deep troughs, and curved grooves. There is a low occurrence of subparallel linear fractures (Figure 4).

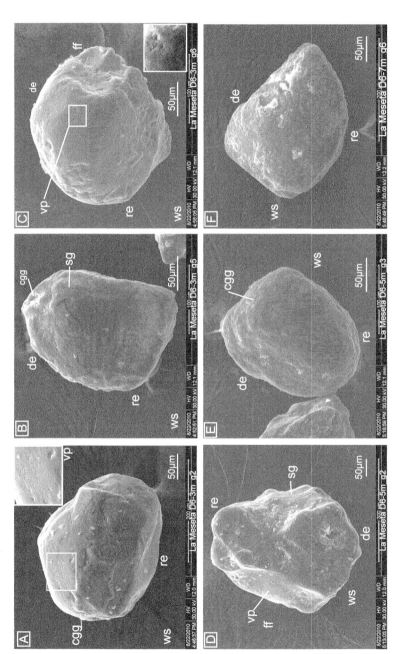

Figure 3. Representative samples from the Eocene La Meseta Formation. (a) Sample D6-3m, a weathered grain that is very rounded with rounded edges (re), weathered surface (ws), weathered crescentic gouges (cgg), and v-shaped impact pits (vp). (b) D6-3m grain with crescentic gouges (cgg), straight grooves (sg), dissolution etching (de), rounded edges (re), and weathered surface (ws). (c) V-shaped impact pits (vp) and fracture faces (ff) on grain D6-3m, along with weathering features, including dissolution etching (de), rounded edges (re), and weathered surface (ws). (d) Sample D6-5m is highly weathered with v-shaped impact pits (vp), fracture faces (ff), straight grooves (sg), dissolution etching (de), rounded edges (re), and weathered surface (ws). (e) This highly weathered grain from sample D6-5m displays many weathering features including dissolution etching (de), rounded edges (re), weathered surface (ws), and weathering features superimposed on crescentic gouges (cgg, arrow). (f) Sample D6-7m is a highly weathered grain characterized by dissolution etching (de), rounded edges (re), and weathered surface (ws).

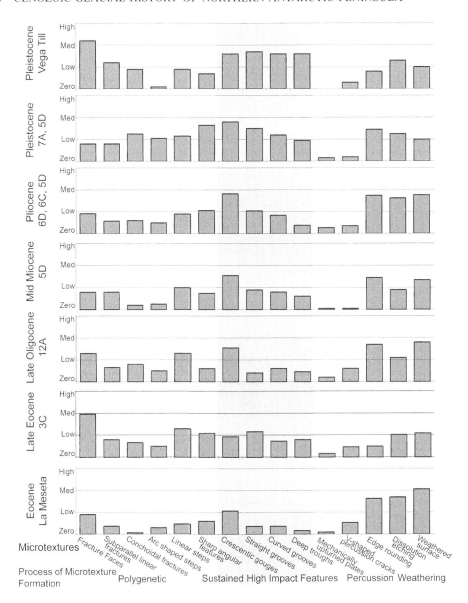

Figure 4. Occurrence of microtextures by age of samples. The *x* axis is 15 identifiable microtextures, and the *y* axis is relative abundance of features based on classification of *Mahaney* [2002]. Feature classification is zero, low, medium, and high abundances. Processes of microtexture formation are from the work of *Sweet and Soreghan* [2010]. Gray area highlights sustained high-impact features formed during glacial transport. Graphs are representative of entire intervals of time and are averaged within units.

4.2. Late Oligocene and Middle Miocene

4.2.1. Late Oligocene Joinville Plateau. Late Oligocene samples from Site 12A-2 (12A-2_67-68cm and 12A-2_225-226cm, Figure 5) contain semiangular grains with rounded edges. Edge rounding is the most common feature, occurring in medium abundance in over half of the grains. Microtextures appear to have a later phase of physical weathering, based on dulling and rounding

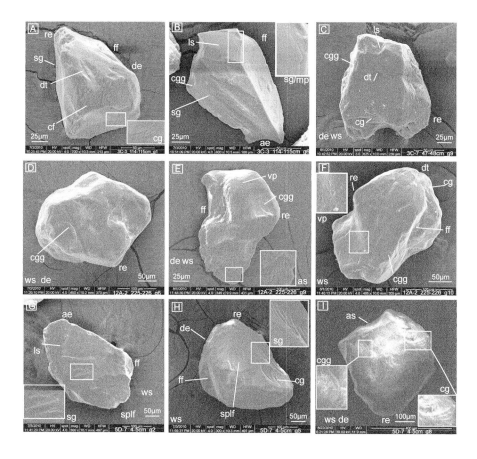

Figure 5. Representative samples from the (top) late Eocene, (middle) late Oligocene, and (bottom) the middle Miocene. (a) Late Eocene sample 3C-3_114-115cm contains straight and curved grooves (sg and cg), deep troughs (dt), fracture faces (ff), and conchoidal fractures (cf). The grain is slightly weathered, showing rounded edges (re) and dissolution etching (de). (b) Extremely fresh and angular late Eocene grain from sample 3C-3_114-115cm with angular edges (ae), fracture faces (ff), linear steps (ls), crescentic gouges (cgg), straight grooves (sg), and mechanically upturned plates (mp). (c) Grain from late Eocene sample 3C-7_47-48cm displays rounded edges (re), weathered surface (ws), dissolution etching (de), deep troughs (dt), crescentic gouges (cgg), curved grooves (cg), and linear steps (ls). (d) Grain from late Oligocene sample 12A-2_225-226cm illustrating crescentic gouges (cgg) that are overprinted by weathering features such as weathered surface (ws), rounded edges (re), and dissolution etching (de). (e) Late Oligocene 12A-2_225-226cm grain showing fracture faces (ff), v-impact pits (vp), arc steps (as), crescentic gouges (cgg), rounded edges (re), dissolution etching (de), and a weathered surface (ws). (f) Late Oligocene sample 12A-2_225-226cm grain illustrating crescentic gouges (cgg), curved grooves (cg), deep troughs (dt), and overprinting by weathering features including rounded edges (re). Grains from this site also contain v-impact pits (vp) and fracture faces (ff). (g) Grain from mid-Miocene sample 5D-7_4-5cm with straight grooves (sg), linear steps (ls), fracture faces (ff), subparallel linear fractures (splf), angular edges (ae), and weathered surface (ws). (h) Grain from mid-Miocene sample 5D-7_4-5cm has straight and curved grooves (sg and cg), subparallel linear fractures (splf), fracture faces (ff), rounded edges (re), dissolution etching (de), and a weathered surface (ws). (i) Grain from mid-Miocene sample 5D-7_4-5cm illustrating weathered surfaces (ws), faint dissolution etching (de), and rounded edges (re), which overprint earlier stage crescentic gouges (cgg), curved grooves (cg), and arc shaped steps (as).

of features and edges. There are low/medium abundances of crescentic gouges and curved grooves, both with rounded feature edges (Figure 4).

4.2.2. Middle Miocene. Middle Miocene samples from Site 5D (5D-6_1-2cm, 5D-7_4-5cm, and 5D-10_44-45cm, Figure 5) are angular to subangular with rounded edges. The most abundant features, in medium abundances, are crescentic gouges, weathered surfaces, and edge rounding. The majority of grains display features of dissolution etching. Other attributes include fracture faces, subparallel linear fractures, linear steps, sharp angular features, straight and curved grooves, and deep troughs. Most of the impact features have subsequently been rounded, resulting in rounded edges and low relief.

4.3. Pliocene

4.3.1. Early Pliocene. The early Pliocene samples from SHALDRIL sites 6C and 5D (6C-6_46-47cm, 5D-2_26-27cm, and 5D-4_27-28cm, Figure 6) consist of a wide range of grain shapes, averaging slightly more than half rounded grains and slightly less than half angular grains. Rounded edges are common and appear in a majority of the grains. Most grains show evidence of silica precipitation, edge rounding, dissolution etching, and weathered surfaces. Crescentic gouges are the most common feature, occurring in medium abundance. There are low abundances of fracture faces, linear steps, sharp angular features, and straight grooves.

4.3.2. Late early Pliocene. Late early Pliocene deposits from SHALDRIL sites 6D and 6C (6D-1_13-14cm and 6C-2_61-62cm, Figure 6) contain a range of grain shapes, with most grains displaying a high degree of angularity and low degree of rounding. The most abundant microtextures are crescentic gouges. There is a low abundance of fracture faces, subparallel linear and conchoidal fractures, linear steps, sharp angular features, and straight and curved grooves. The features are commonly very fresh, with individual grains having sharp edges.

4.4. Pleistocene

4.4.1. Offshore Joinville Plateau. The Pleistocene samples from SHALDRIL sites 5D and 7A (7A_1-2cm and 5D-1_10-11cm, Figure 6) display a broad scope of grain shapes with a wide range of length to width ratios. The grain edges range in degree of rounding from rounded to quite angular. The most abundant microtextures (medium abundance) include crescentic gouges, sharp angular features, and straight and curved grooves. There is a low abundance of fracture faces, subparallel linear and conchoidal fractures, arc-shaped and linear steps, curved grooves, and deep troughs.

4.4.2. Late Pleistocene till. The Pleistocene till from piston core NBP02-01-PC06 (70–72cm, Figure 6) contains mostly highly angular grains, with individual grains having a high aspect ratio. Edges vary between fresh to weathered and rounded. There is a medium abundance of straight and curved grooves, fracture faces, crescentic gouges, and deep troughs on the majority of grains. There is a low frequency of weathered surfaces, dissolution etching, and edge rounding.

5. DISCUSSION

Sustained high-impact features are created through extreme shear stress during glacial transport and include crescentic gouges, straight and curved grooves, deep troughs and mechanically upturned plates [*Mahaney et al.*, 1996; *Sweet and Soreghan*, 2010], (Figures 3, 5, 6). In this

Figure 6. Samples from (top) Pleistocene till, from Joinville Plateau, (middle) sites 7A and 5D, and (bottom) Pliocene. (a) Sample NBP02-01_PC-06_70-72cm displays curved grooves (cg), straight grooves (sg), crescentic gouges (cgg), deep troughs (dt), and subparallel linear fractures (splf) and is fresh with little/no weathering and angular edges (ae). (b) Grain from Pleistocene sample NBP02-01_PC-06_70-72cm illustrating crescentic gouges (cgg), curved grooves (cg), fracture faces (ff), and linear steps (ls). (c) Pleistocene NBP02-01_PC-06_70-72cm grain with deep troughs (dt), curved grooves (cg), rounded edges (re), dissolution etching (de), sharp angular features (saf), and a weathered surface (ws). (d) Grain from Pleistocene SHALDRIL sample 7A_1-2cm showing rounded edges (re), abundant curved grooves (cg), straight grooves (sg), and arc shaped steps (as). (e) Grain from Pleistocene sample 5D-1_10-11cm is very fresh with angular edges (ae), straight grooves (sg), deep troughs (dt), and linear steps (ls). (f) Sample 5D-1_10-11cm contains deep troughs (dt), curved grooves (cg), crescentic gouges (cgg), v-impact pits (vp), conchoidal fractures (cf), arc steps (as), dissolution etching (de), and rounded edges (re). (g) Grain from Pliocene sample 6D-1_13-14cm with rounded edges (re), a weathered surface (ws), and dissolution etching (de). The grain also contains fracture faces (ff), straight grooves (sg), crescentic gouges (cgg), arc steps (as), linear steps (ls), mechanically upturned plates (mp), and subparallel linear features (splf). (h) Grains from early Pliocene sample 5D-2_26-27cm have rounded edges (re), a very weathered surface (ws), relict straight grooves (sg), and weathered crescentic gouges (cgg). (i) This Pliocene 5D-4_27-28cm sample is extremely fresh with angular edges (ae), crescentic gouges (cgg), curved and straight grooves (cg and sg), sharp angular features (saf), and subparallel linear features (splf).

study, Pleistocene tills provide a baseline for glacially transported grains. Generally, the features observed most commonly in tills, in order of abundance, are (1) subparallel linear and conchoidal fractures; (2) sharp angular features, straight grooves, and deep troughs; and (3) curved grooves, low relief crescentic gouges, arc steps, linear steps, mechanically upturned plates, lattice shatters, and fracture features. The relative abundance of subparallel linear fractures can be related to the amount of stress incurred during transport. Smaller glaciers and ice caps transport sediment in a lower stress regime and impart a lower relative abundance of subparallel linear fractures as opposed to large ice sheets which transport grains with higher stresses and have higher abundances of subparallel linear fractures [*Mahaney*, 2002].

The oldest sediments examined for this study are from the Eocene La Meseta Formation, which historically has been interpreted as a nonglacial deposit, void of any ice-rafted debris or any other

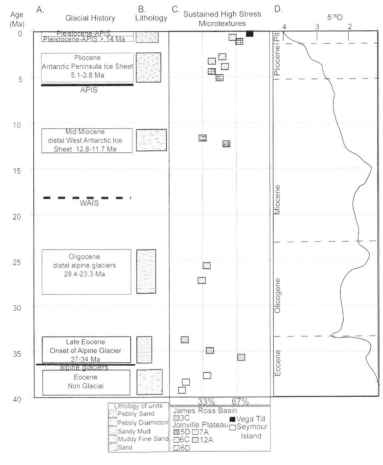

Figure 7. Overall stratigraphy, grain surface textures, and glacial history of the James Ross Basin. (a) Age constraints on glacial evolution of the West Antarctic Ice Sheet (WAIS) and Antarctic Peninsula Ice Sheet (APIS) [*Anderson et al.*, 2011]. (b) Lithostratigraphic column of units sampled. (c) Plot of frequency of sustained high-stress microtextures based on the work of *Sweet and Soreghan* [2010]. Frequency of particular features was calculated over the entire sample interval using the presence or absence of that feature and taking the mean. The relative frequency of sustained high-stress microtextures was calculated by averaging all five high-stress features for the specific sample. (d) Modified $\delta^{18}O$ curve from the work of *Zachos et al.* [2001].

evidence for glaciation [e.g., *Marenssie et al.*, 2002], (Figure 7). This interpretation is supported by this investigation, which has yielded no compelling evidence for glaciation at this time. The samples examined from the La Meseta Formation contain a low abundance of crescentic gouges and no other sustained high-stress microtextures (Figures 3 and 4). Thus, the La Meseta Formation provides a preglacial baseline with which to compare samples from younger strata in the region.

SHALDRIL cores that sampled the late Eocene (Site 3C) consist of poorly sorted muddy/silty very fine sand with few subangular-subrounded pebbles. The age of the sampled interval has been constrained to 37–34 Ma based on diatom biostratigraphy and Sr dating [*Bohaty et al.*, this volume]. The samples from this core are characterized by a medium abundance of sustained high-impact microtextures, in contrast to the Seymour Island samples which are relatively void of such features (Figure 5). The samples also contain zero to low abundance of subparallel linear fractures and are characterized by high angularity. The dramatic difference in sand grain surface textures between the Eocene La Meseta samples and the SHALDRIL Site 3C (late Eocene) samples indicates that the onset of glaciation in the northern Antarctic Peninsula occurred at the end of the Eocene, likely restricted to alpine glaciation. This is based on sharp-edged microtextures and low abundance of subparallel linear fractures [*Mahaney*, 2002]. These findings are consistent with palynological results, which indicate a significant change in the vegetation of the area at this time [*Warny and Askin*, this volume(a), this volume(b)] and seismic stratigraphic analyses that show no evidence of ice having grounded on the continental shelf [*Smith and Anderson*, 2010]. This corresponds to the beginning of a large positive oxygen isotopic excursion [*Zachos et al.*, 2001] and fall in sea level [*Miller et al.*, 2005] marking the transition from greenhouse to icehouse conditions (Figure 7).

The upper Oligocene deposits sampled by SHALDRIL Site 12A consist of dark gray sandy mud to muddy sand that contains clay lenses, burrows, and rare dropstones. This deposit is interpreted as representing a distal deltaic or shoreface environment [*Wellner et al.*, this volume]. Sand grains from the upper Oligocene display a presence, but a lower abundance, of glacially derived microtextures compared to Site 3C deposits, along with a high degree of physical weathering. A lower abundance of glacially derived microtextures may be a function of the distal location of this site in comparison to Site 3C. This is consistent with palynological evidence for a colder climate during the Oligocene than existed in the Eocene [*Warny and Askin*, this volume(a), this volume(b)].

The middle Miocene deposits recovered at SHALDRIL Site 5D sampled pebbly gray diamicton, interpreted as glacimarine sediment [*Wellner et al.*, this volume] and thus indicate more extreme glacial conditions than previously existed in the region. The palynomorphs present at this site also indicate colder conditions relative to the late Oligocene, with limited land taxa and dominantly sea-ice species offshore [*Warny and Askin*, this volume(a), this volume(b)]. The microtextures of grains from Site 5D show an increased abundance of high-impact features relative to the late Oligocene samples from Site 12A (Figures 4 and 5). Individual grains display sustained high-impact features, but have a high degree of rounding (Figure 7). These glacial grains have medium to high abundances of physical weathering features and were likely transported by contour currents after being delivered to the drill site by icebergs drifting under the influence of the Weddell Gyre. This is consistent with inferred expansion of the West Antarctic Ice Sheet (WAIS) onto the southwestern Weddell Sea continental shelf in the middle Miocene [*Smith and Anderson*, 2010].

The early Pliocene section recovered at sites 6C and 6D consists of pebbly glacimarine sediments. The sand grains from these sediments contain a mixture of well-rounded contour current-transported grains and glacially transported grains. The oldest deposits from Site 6C include very dark greenish gray, medium to coarse sand, and poorly sorted diamicton. Sand grains show a clear increase in the abundance of sustained high-stress microtextures (Figure 4), consistent with other evidence for extreme glacial conditions at this time, culminating in initial advance of the Antarctic Peninsula Ice Sheet (APIS) onto the Joinville Plateau [*Smith and Anderson*, this volume].

Pleistocene samples contain the highest abundance of sustained high-impact surface features (Figure 7), which is consistent with evidence for an expanding and contracting APIS during the Pleistocene. The SHALDRIL Pleistocene samples are only slightly lower in abundance of sustained high-impact features relative to the late Pleistocene till samples from piston cores.

6. CONCLUSION

During the late Eocene, there was onset of alpine glaciation in the northern Antarctic Peninsula. The occurrence of glacial surface features records the continued presence of alpine glaciers during the late Oligocene, which is consistent with palynological evidence for significantly cooler conditions relative to the late Eocene.

Samples from mid-Miocene deposits include ice-rafted grains from distant sources, likely West Antarctica, that were probably transported to the study area by icebergs drifting under the influence of the Weddell Gyre. Samples from early Pliocene strata show an increase in the abundance of high-impact features and record the early development of the APIS. Pleistocene till samples display the highest abundance of glacially influenced microtextures and record the extreme glacial conditions that have characterized the region since the early Pliocene.

Acknowledgments. This work was supported by the National Science Foundation grant NSF-OPP-0125922 to J.B.A. The authors extend their appreciation to those who provided samples for this study: Anne Grunnow and Larry Krissek (Byrd Polar Research Center at Ohio State University), Alexander Simms (University of California, Santa Barbara), and Steven Petrushak (Antarctic Marine Research Facility). We thank Julia Wellner for constructive discussion and encouragement of work. A special thank you to Richard Thibault, Alvin Orbaek, Hugh Daigle, and Kyusei Tsuno for help with sample preparation and scanning electron microscopy training. This manuscript was greatly improved by reviews from Kari Strand (University of Oulu, Finland) and Gerilyn Soreghan (University of Oklahoma).

REFERENCES

Anderson, J. B., et al. (2011), Progressive Cenozoic cooling and the demise of Antarctica's last refugium, *Proc. Natl. Acad. Sci. U. S. A.*, *108*(28), 11,356–11,360, doi:10.1073/pnas.1014885108.

Bohaty, S. M., D. K. Kulhanek, S. W. Wise Jr., K. Jemison, S. Warny, and C. Sjunneskog (2011), Age assessment of Eocene–Pliocene drill cores recovered during the SHALDRIL II expedition, Antarctic Peninsula, in *Tectonic, Climatic, and Cryospheric Evolution of the Antarctic Peninsula*, doi: 10.1029/2010SP001049, this volume.

Bull, P. A. (1981), Environmental reconstruction by electron microscopy, *Prog. Phys. Geogr.*, *5*(3), 368–397.

Heroy, D. C., and J. B. Anderson (2005), Ice-sheet extent of the Antarctic Peninsula region during the Last Glacial Maximum (LGM)—Insights from glacial geomorphology, *Geol. Soc. Am. Bull.*, *117*(11–12), 1497–1512.

Krinsley, D., and S. Margolis (1969), A study of quartz sand grain surface textures with scanning electron microscope, *Trans. N. Y. Acad. Sci.*, *31*(5), 457–477.

Krinsley, D. H., and J. C. Doornkamp (1973), *Atlas of Quartz Sand Surface Textures*, 91 pp., Cambridge Univ. Press, Cambridge, U. K.

Mahaney, W. C. (2002), *Atlas of Sand Grain Surface Textures and Applications*, 237 pp., Oxford Univ. Press, Oxford, U. K.

Mahaney, W. C., and V. Kalm (1995), Scanning electron-microscopy of Pleistocene tills in Estonia, *Boreas*, *24*(1), 13–29.

Mahaney, W. C., and V. Kalm (2000), Comparative scanning electron microscopy study of oriented till blocks, glacial grains and Devonian sands in Estonia and Latvia, *Boreas*, *29*(1), 35–51.

Mahaney, W. C., W. Vortisch, and P. J. Julig (1988), Relative differences between glacially crushed quartz transported by mountain and continental ice; some examples from North America and East Africa, *Am. J. Sci.*, *288*(8), 810–826.

Mahaney, W. C., G. Claridge, and I. Campbell (1996), Microtextures on quartz grains in tills from Antarctica, *Palaeogeogr. Palaeoclimatol. Palaeoecol.*, *121*(1–2), 89–103.

Maldonado, A., et al. (2005), Miocene to Recent contourite drifts development in the northern Weddell Sea (Antarctica), *Global Planet. Change*, *45*(1–3), 99–129.

Marenssi, S. A., L. I. Net, and S. N. Santillana (2002), Provenance, environmental and paleogeographic controls on sandstone composition in an incised-valley system: The Eocene La Meseta Formation, Seymour Island, Antarctica, *Sediment. Geol.*, *150*(3–4), 301–321.

Marshall, J. R. (1987), *Clastic Particles: Scanning Electron Microscopy and Shape Analysis of Sedimentary and Volcanic Clasts*, 346 pp., Van Nostrand Reinhold, New York.

Miller, K. G., M. A. Kominz, J. V. Browning, J. D. Wright, G. S. Mountain, M. E. Katz, P. J. Sugarman, B. S. Cramer, N. Christie-Blick, and S. F. Pekar (2005), The phanerozoic record of global sea-level change, *Science*, *310*(5752), 1293–1298.

Smith, R. T., and J. B. Anderson (2010), Ice-sheet evolution in James Ross Basin, Weddell Sea margin of the Antarctic Peninsula: The seismic stratigraphic record, *Geol. Soc. Am. Bull.*, *122*(5–6), 830–842.

Smith, R. T., and J. B. Anderson (2011), Seismic stratigraphy of the Joinville Plateau: Implications for regional climate evolution, in *Tectonic, Climatic, and Cryospheric Evolution of the Antarctic Peninsula*, doi: 10.1029/2010SP000980, this volume.

Smith, W. H. F., and D. T. Sandwell (1997), Global seafloor topography from satellite altimetry and ship depth soundings, *Science*, *277*, 195–196.

Stickley, C. E., K. St John, N. Koç, R. W. Jordan, S. Passchier, R. B. Pearce, and L. E. Kearns (2009), Evidence for middle Eocene Arctic sea ice from diatoms and ice-rafted debris, *Nature*, *460*(7253), 376–388.

Sweet, D. E., and G. S. Soreghan (2010), Application of quartz sand microtextural analysis to infer cold-climate weathering for the equatorial Fountain Formation (Pennsylvanian-Permian, Colorado, USA), *J. Sediment. Res.*, *80*(7–8), 666–677.

Warny, S., and R. Askin (2011a), Vegetation and organic-walled phytoplankton at the end of the Antarctic greenhouse world: Latest Eocene cooling events, in *Tectonic, Climatic, and Cryospheric Evolution of the Antarctic Peninsula*, doi: 10.1029/2010SP000965, this volume.

Warny, S., and R. Askin (2011b), Last remnants of Cenozoic vegetation and organic-walled phytoplankton in the Antarctic Peninsula's icehouse world, in *Tectonic, Climatic, and Cryospheric Evolution of the Antarctic Peninsula*, doi: 10.1029/2010SP000996, this volume.

Wellner, J. S., J. B. Anderson, W. Ehrmann, F. M. Weaver, A. Kirshner, D. Livsey, and A. Simms (2011), History of an evolving ice sheet as recorded in SHALDRIL cores from the northwestern Weddell Sea, Antarctica, in *Tectonic, Climatic, and Cryospheric Evolution of the Antarctic Peninsula*, doi: 10.1029/2010SP001047, this volume.

Zachos, J., M. Pagani, L. Sloan, E. Thomas, and K. Billups (2001), Trends, rhythms, and aberrations in global climate 65 Ma to present, *Science*, *292*(5517), 686–693.

J. B. Anderson and A. E. Kirshner, Department of Earth Science, Rice University, 6100 S. Main, MS-126, Houston, TX 77005, USA. (kirshner@rice.edu)

Last Remnants of Cenozoic Vegetation and Organic-Walled Phytoplankton in the Antarctic Peninsula's Icehouse World

Sophie Warny

Department of Geology and Geophysics and Museum of Natural Science, Louisiana State University
Baton Rouge, Louisiana, USA

Rosemary Askin

Jackson, Wyoming, USA

A late Oligocene, a middle Miocene, and two adjacent Pliocene sections were sampled off the coast of the Antarctic Peninsula in shelf sediments on the Joinville Plateau, Weddell Sea. Drilling was conducted from the research vessel icebreaker *Nathaniel B. Palmer* during the 2006 SHALDRIL campaign. The drill holes recovered sediment cores that each span a short interval of time because of extensive sea ice constraints during drilling. Despite this limitation, the palynomorphs extracted from these sediments help constrain the region's past environmental conditions during three periods of the "icehouse" world and confirm that tundra vegetation persisted in the Antarctic Peninsula up to at least 12.8 Ma. The terrestrial palynological data reflect southern beech and conifer-dominated woodlands and tundra during the Oligocene, with reduction to pockets of tundra with probably stunted beech and podocarp conifers by the middle Miocene. During both the Oligocene and the Miocene, the phytoplankton were dominated by small sea ice-tolerant opportunistic species taking advantage of the migration of most dinoflagellate cysts to more hospitable parts of the ocean. By the Pliocene, only limited pockets of vegetation may have existed.

1. INTRODUCTION

Understanding the timing and characteristics of the evolving post-Eocene Antarctic Ice Sheet is fundamental for constraining the mechanisms driving climate cooling. In turn, this information is needed by modelers quantifying how ice sheet expansion and contraction impacts global climate. The Antarctic Peninsula is located in lower latitudes compared to the rest of the Antarctic

Tectonic, Climatic, and Cryospheric Evolution of the Antarctic Peninsula
Special Publication 063
Copyright 2011 by the American Geophysical Union.
10.1029/2010SP000996

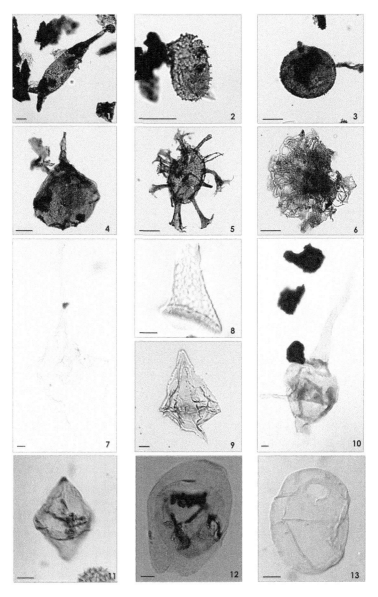

Figure 1. Light photomicrographs of key species of reworked dinoflagellate cysts from the Mesozoic. Scale bar is 20 um. For each image, the numbers indicate the taxon name, sample number (e.g. NBP0602A-5D-10), the sample depth (e.g., 12), and the England Finder coordinates: 1, *Batiolodinium radiculatum*, NBP0602A-5D-10.12, M35/1; 2, *Prolixosphaeridium inequiornatum*, NBP0602A-5D-10.12, Q43/2; 3, *Apteodinium punctatum*, NBP0602A-5D-10.12, T39/1; 4, *Cribroperidinium longicorne*, NBP0602A-5D-10.12, R31/3; 5, *Oligosphaeridium pulcherinum*, NBP0602A-5D-10.12, R17/4; 6, ?*Glaphyrocysta* sp., NBP0602A-5D-10.12, P43/2; 7, *Muderongia tetracantha*, NBP0602A-5D-8.37, F32/4; 8, fragment of *Odontochitina porifera* operculum, NBP0602A-5D-11.64, F24/2; 9, *Spinidinium* cf. *lanternum*, NBP0602A-5D-7.27, T28/1; 10, *Odontochitina operculata*, NBP0602A-12A-2.232, M15/1; 11, *Diconodinium pussilum*, NBP0602A-12A-2.232, N35/1; 12, *Manumiella druggii*, NBP0602A-5D-11.313, W16/3; 13, *Cyclopsiella* sp., NBP0602A-5D-11.313, H20/3.

continent and is assumed to have been the last vegetated region to succumb to the encroaching ice sheet. The timing of ice sheet inception in the Antarctic Peninsula has been controversial. *Bart and Anderson* [1995], based on seismic stratigraphic evidence, argued that the ice sheet expanded onto the continental shelf on the Pacific side of the Antarctic Peninsula as early as the middle Miocene. *Bart et al.* [2005, 2007] further demonstrated that at least 13 Antarctic Peninsula Ice Sheet (APIS) grounding events occurred during the late Miocene/early Pliocene, represented at various shelf sites (between 7.94 and 5.12 Ma). *Birkenmajer* [1996] and *Dingle and Lavelle* [1998] found evidence for glacial deposits in the King George Island region as early as the Oligocene. Little is known of this Oligocene ice. What were its characteristics and extent? Was it a

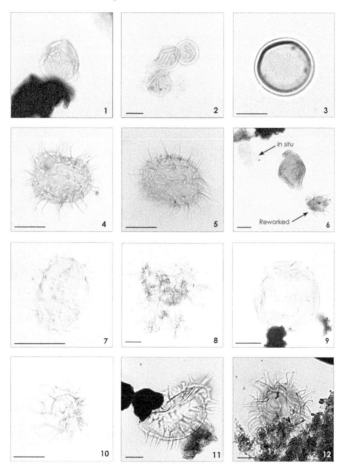

Figure 2. Light photomicrographs of key species of dinoflagellate cysts, acritarchs, and freshwater algae from the Cenozoic. Scale bar is ~20 um. Numbers represent the following: 1, zygospore of Zygnemaceae, NBP0602A-5D-11.230, S18/4; 2, thin-walled small leiosphere, NBP0602A-12A-1.30, S24/1; 3, thick-walled leiosphere, NBP0602A-5D-10.86, P32/4; 4, *Micrhystridium* sp., NBP0602A-5D-10.171, V46/1; 5, *Micrhystridium* sp., NBP0602A-5D-10.171, T23/1; 6, difference in maturation index between reworked versus in situ *Micrhystridium* sp., NBP0602A-5D-11.230, S31/2; 7, unknown small chorate dinoflagellate cyst, most likely in situ, NBP0602A-5D-11.151, E19/2; 8, *Glaphyrocysta ordinata*, NBP0602A-5D-11.151, E13/1; 9, *Batiacasphaera* sp., NBP0602A-5D-11.151, K28/1; 10, unknown small chorate dinoflagellate cyst, most likely in situ NBP0602A-5D-11.64, F13/3; 11, *Operculodinium centrocarpum*, NBP0602A-5D-7.27, S33/2; 12, *Spiniferites* sp., possibly in situ, NBP0602A-5D-7.27, T40/2.

localized ice sheet or glacier? What type of vegetation survived through the Oligocene, where, and how far through the Miocene were areas of the Antarctic Peninsula vegetated? What is the age of the last record of vegetation in the northern tip of the Antarctic Peninsula?

To help answer some of these questions, a palynological analysis of three key time intervals (late Oligocene, middle Miocene, and early Pliocene) was undertaken. The main goal of this study was to better constrain the evolution of climate and environmental conditions around the Joinville Plateau during the post-Eocene "icehouse" world. The palynological data collected from SHALDRIL NBP0602A provide excellent information on the regional vegetation and phyto-plankton communities during these time intervals. The results are presented below, organized by time interval sampled, from Oligocene to Pliocene. Some of the key species of acritarchs and

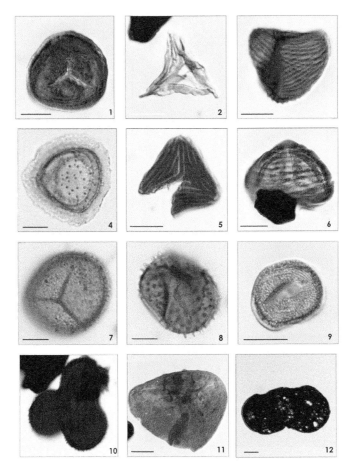

Figure 3. Light photomicrographs of key species of pollen and spores reworked from the Permian and Mesozoic. Scale bar is ~20 um. Numbers represent the following: 1, *Cyatheacidites annulatus*, NPB0602A-5D-7.27, L32/0; 2, *Clavifera triplex*, NBP0602A-5D-10.171, K40/2; 3, *Cicatricosisporites australiensis*, NBP0602A-5D-11.151, E39/1; 4, *Aequitriradites spinulosus*, NBP0602A-12A-2.262, J46/2; 5, *Cicatrico-sisporites* sp. cf. *Cuvierina ludbrooki*, NBP0602A-5D-10.171, K23/1; 6, *Contignisporites cooksonii*, NBP0602A-5D-12.83, G31/0; 7, *Baculatisporites comaumensis*, NBP0602A-12A-1.62, U43/2; 8, *Ceratos-porites* cf. *equalis*, NBP0602A-5D-10.171, U33/4; 9, *Classopollis classoides*, NBP0602A-5D-10.12, G23/4; 10, *Trilobosporites trioreticulosus*, NBP06-02A-5D-10.171, P38/4; 11, *Granulatisporites trisinus*, NBP0602A-12A-1.30, S28/4; 12. Bisaccate pollen, possibly taeniate, NBP0602A-5D-9.43, D32/3.

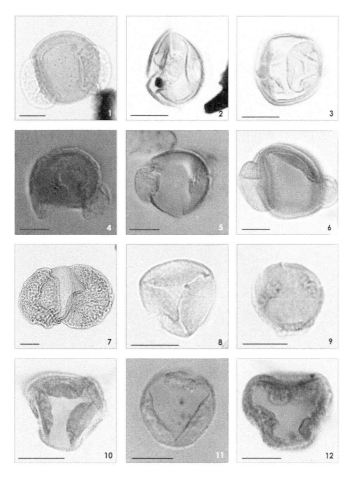

Figure 4. Light photomicrographs of key species of podocarp conifer pollen from the Cenozoic, including reworked and in situ specimens. Scale bar is ~20 um. Numbers represent the following: 1, *Podocarpidites* sp. cf. *Podocarpidites exiguus*, NBP0602A-12A-2.69, R31/1; 2, *Phyllocladidites* sp.*a*., NBP0602A-12A-2.232, O29/0; 3, *Phyllocladidites mawsonii* (reworked), NBP0602A-12A-2.171, K34/0; 4, *Phyllocladidites* sp.*a*, NBP0602A-5D-10.171, D12/2; 5, *Phyllocladidites* sp.*a*, NBP0602A-5D-10.171, F38/3; 6, *Phyllocladidites* sp.*a*, NBP0602A-12A-2.144, K44/2; 7, *Podocarpidites* sp., NBP0602A-5D-7.27, R34/2; 8–12, *Podosporites microsaccatus-Trichotomosulcites subgranulatus* complex; 8, NBP0602A-12A-2.202, V38/1; 9, NBP0602A-5D-11.64, N37/4; 10, NBP0602A-12A-1.62, V26/3; 11, NBP0602A-12A-2.69, F26/0; 12, NBP0602A-12A-1.95, T22/3.

dinoflagellate cysts discussed in the text are illustrated in Figures 1 and 2. The main pollen and spores are illustrated in Figures 3, 4, 5, and 6.

2. OLIGOCENE SAMPLED BETWEEN 28.6 AND 24.0 MA (CORE NBP0602A-12A)

2.1. Material and Methods

Borehole NBP0602A-12A was drilled at 63°16.354′S latitude and 52°49.501′W longitude (Figure 7) and sampled the strata offshore and down dip of James Ross Island on the Joinville

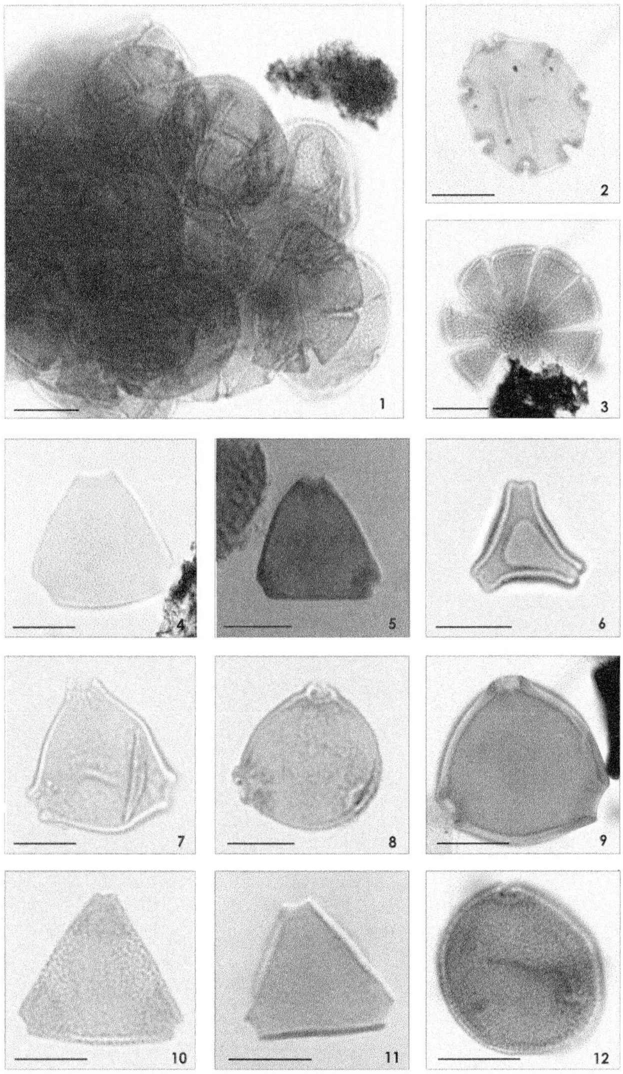

Figure 5. Light photomicrographs of key species of pollen from the Cenozoic sections, including reworked and in situ specimens. Scale bar is ~20 um. Numbers represent the following: 1, dissemination cluster of *Nothofagidites lachlaniae*, NBP0602A-5D-11.313, U26/1; (this is a solid evidence that these are in place in the Miocene); 2, *Nothofagidites* sp. cf. *Nothofagidites flemingii*, NBP0602A-12A-1.62, S23/3; 3, *Polycolpites* sp. NBP0602A-5D-11.313, M42/2; 4, *Proteacidites* sp. cf. *Proteacidites parvus*, NBP0602A-5D-13.63, D24/0; 5, *Propylipollis* sp. NBP0602A-12A-2.232, M42/0; 6, ?*Proteacidites* sp., possibly missing sexine, NBP0602A-12A-2.171, R44/2; 7, ?*Triporopollenites* sp., NBP0602A-5D-11.230, U13/1; 8, *Triporopollenites* sp., NBP0602A-12A-2.69, N24/2; 9, *Triporopollenites* sp., NBP0602A-5D-11.313, M24/1; 10, *Propylipollis reticuloscabratus*, NBP0602A-5D-11.230, V40/3; 11, *P. parvus*, NBP0602A-12A-2.202, K11/1; 12, triporate reticulate unknown, NBP0602A-5D-11.313, O25/3.

Plateau, Weddell Sea (Figure 8). This hole was drilled in 442 m of water depth and is composed of a series of three drilled sections or cores (1R$_a$, 2R$_a$, and 3R$_a$) with intervals of low to no recovery (Figure 9) (Shipboard Scientific Party, SHALDRIL II 2006 NBP0602A cruise report, 2006,

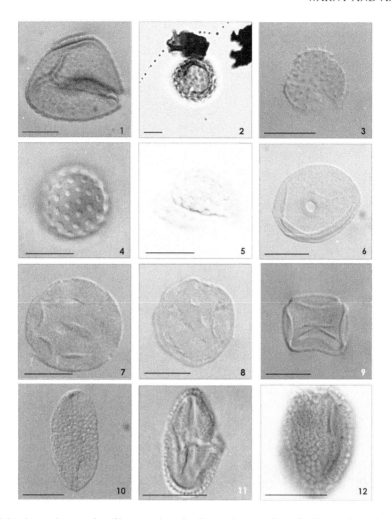

Figure 6. Light photomicrographs of key species of pollen and spores from the Cenozoic tundra (see text for explanation). Scale bar is ~20 um. Numbers represent the following: 1, *Coptospora* sp., NBP0602A-5D-10.171, E32/1; 2, Asteraceae, *Tubulifloridites* sp. NBP0602A-5D-10.86, T39/1; 3, Asteraceae, NBP0602A-5D-10.12, F46/3; 4, *Chenopodipollis* sp., NBP0602A-12A-2.103, F38/2; 5, *Chenopodipollis* sp., NBP0602A-12A-2.103, G34/0; 6, Poaceae, NBP0602A-5D-8.37, H41/0; 7, Caryophyllaceae (*Colobanthus*-type), NBP0602A-5D-11.230, H37/0; 8, Caryophyllaceae (*Colobanthus*-type), NBP0602A-5D-7.27, L42/2; 9, tetraporate unknown, cf. *Haloragacidites* sp., NBP0602A-5D-10.171, N21/3; 10, *Liliacidites* sp. (possibly reworked), NBP0602A-5D-10.171, N21/4; 11, *Rhoipites* sp., NBP0602A-5D-11.230, H44/3; 12, *Rhoipites* sp., NBP0602A-5D-10.171, Q37/2.

available at http://www.arf.fsu.edu/projects/documents/Shaldril_2_Report.pdf, hereinafter Shipboard Scientific Party, 2006). Continuous coring was made impossible by harsh weather conditions and fast moving sea ice at the time of drilling. There was no recovery from the lowest core unit (core $3R_a$), and it is not discussed further here. Age assignment is based on diatom biostratigraphy. Units in core $2R_a$ (referred to as NBP0602A-12A-2) and most of core $1R_a$ (referred to as NBP0602A-12A-1) are assigned a maximum age of 28.6 Ma and a minimum age of 24.0 Ma, based on the occurrence of several key diatoms, including *Cavitatus jouseanus*,

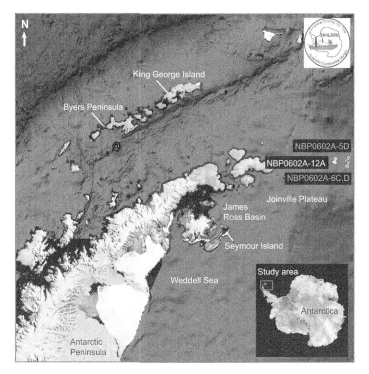

Figure 7. Location of SHALDRIL NBP0602A-12A (63°16.354′S latitude and 52°49.501′W longitude), NBP0602A-5D (63°15.090′S latitude and 52°21.939′W longitude), and NBP0602A-6C/6D (63°20.268′S latitude and 52°22.032′W longitude) drill holes on the Joinville Plateau, Weddell Sea, Antarctica. Location based on coordinates provided by Shipboard Scientific Party (2006). Satellite image from Google Earth and NASA.

Cavitatus rectus, Kisseleviella cicatricata, and *Kisseleviella tricoronata* [*Bohaty et al.,* this volume]. About 4 m of cored section was recovered, including sediment composed of sandy mud to muddy sand. Twelve samples were selected for palynological processing (Figure 9) and processed using a standard palynological technique suited for Antarctic palynology [*Warny and Askin,* this volume].

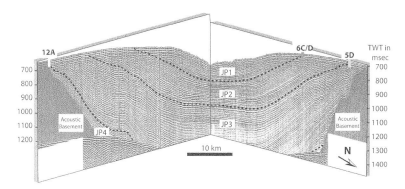

Figure 8. Seismic sections across the Joinville platform and slope showing main stratigraphic units (JP1, JP2, JP3, and JP4) and locations of SHALDRIL sites 5D, 6C/D, and 12A. Modified from *Anderson et al.* [2011].

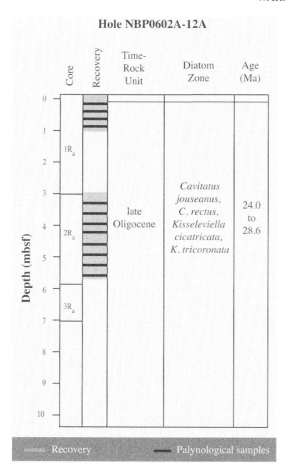

Figure 9. Palynological sample location in relation to the diatom zonal assignments [after *Boharty et al.*, this volume] and age interpretation for borehole NBP0602A-12A (after Shipboard Scientific Party, 2006).

2.2. Palynological Results

2.2.1. General evaluation. Figure 10 presents a summary of palynological results obtained for the late Oligocene sections sampled in NBP0602A-12A-1 and NBP0602A-12A-2. In general, the recovery was low, with concentrations in palynomorphs never exceeding 2000 palynomorphs per gram of dried sediments, in contrast to concentrations up to 120,000 palynomorphs per gram of dried sediments in the nearby late Eocene hole (NBP0602A-3C) [*Warny and Askin*, this volume]. This implies significant climatic deterioration both on land and in the ocean since the end of the Eocene, with greatly decreased numbers of plants and microplankton that produced the palynomorphs. As in the upper SHALDRIL NBP0602A-3C cores [*Warny and Askin*, this volume], the terrestrial assemblages throughout NBP0602A-12A (and these comments also apply to NBP0602A-5D samples discussed below) are dominated by reworked specimens, especially in the younger samples. As with the NBP0602A-3C assemblages, it was not always easy to differentiate between penecontemporaneous and reworked specimens. Counts included an "obvious reworked" category, where morphology and preservation distinguished eroded and redeposited components. As evidenced by the very light color

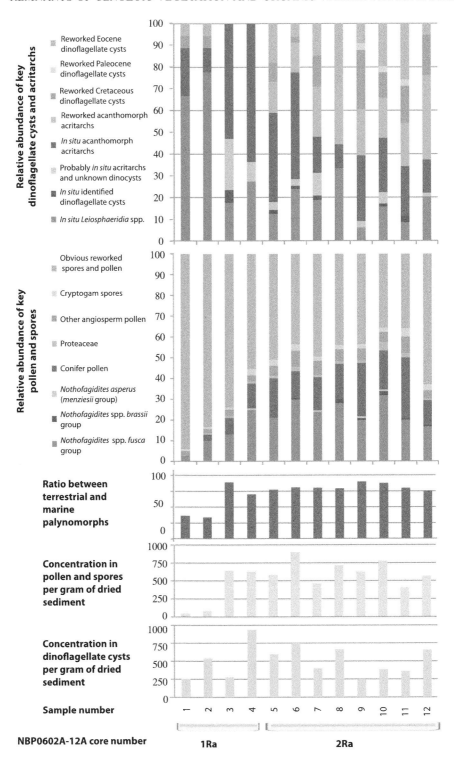

Figure 10. Relative abundance of key palynomorphs, concentration in palynomorphs per gram of dry sediment, and ratio between terrestrial (gray bars) and marine palynomorphs in core NBP0602A-12A.

of some of the Late Cretaceous reworked dinoflagellate cysts, thermal maturity by itself cannot be used as an indicator of reworking. Some pollen and spores are broken, very abraded, and/or corroded, and these battered specimens were thus included in the reworked count. Counts of the terrestrial taxa that might represent part of the Oligocene and Miocene vegetation have erred on the side of inclusion in the penecontemporanous categories, as these specimens are indistinguishable in terms of their morphology and preservation. For instance, counts of the *Nothofagidites* spp. (southern beech pollen), podocarp conifers, Proteaceae and other angiosperms, and cryptogams may be inflated.

2.2.2. Dinoflagellate cysts and acritarchs. The marine assemblage is mainly composed of *Leiosphaeridia* spp., a group of sphaeromorph acritarchs known to predominate in environments with sea ice [*Mudie*, 1992; *Warny et al.*, 2006], and acanthomorph acritarchs (mainly *Micrhystridium* spp.). Other than these two groups of palynomorphs considered as in situ, a few rare dinoflagellate cysts of *Spiniferites* sp. and *Operculodinium* sp. are considered likely to be in place, along with several species of the acritarch *Cymatiosphaera*. In addition to these penecontemporaneous species, the assemblage is largely composed of various reworked specimens of Cretaceous, Paleocene, and Eocene age. Cretaceous and Eocene reworked specimens dominate. Specimens grouped as reworked Cretaceous species include *Palaeoperidinium cretaceum* (Aptian-Albian), *Oligosphaeridium pulcherrimum* (Albian), *Diconodinium cristatum* (Aptian-Cenomanian), *Cyclonephelium compactum* (Albian-Santonian), *Odontochitina operculata* (Aptian-Senonian), *Odontochitina porifera* (Senonian), *Chatangiella victoriensis* (Turonian-Santonian), *Odontochitina spinosa* (Maastrichtian), *Manumiella seymourensis* (Maastrichtian), *Manumiella druggii* (a K/T species, mainly late Maastrichtian but ranging into basal Danian), and the acritarch *Veryhachium reductum* (Cretaceous). These species are mainly of Albian to Maastrichtian age. Reworked Paleocene species are rare; they include a few specimens of *Palaeocystodinium australianum* (late Maastrichtian-Paleocene) and specimens listed as *Spinidinium* cf. *lanternum* that were found in the Paleocene uppermost Sobral Formation on Seymour Island [*Askin*, 1988]. The specimens that are grouped herein as reworked Eocene species were identified as such mainly because they were commonly found as an in situ assemblage in the NBP0602A-3C SHALDRIL core [*Warny and Askin*, this volume] and in the Seymour Island Eocene [e.g., *Wrenn and Hart*, 1988]. The species grouped as Eocene are *Octodinium askiniae* (Eocene), *Vozzhennikovia apertura* (Paleocene-Oligocene), *Enneadocysta partridgei* (middle Eocene to early Oligocene), *Spinidinium macmurdoense* (early Tertiary), *Deflandrea antarctica* (Eocene), and *Deflandrea cygniformis* (early Tertiary).

2.2.3. Pollen and spores. The late Oligocene NBP0602A-12A penecontemporaneous assemblages are dominated by pollen of southern beech, especially the *Nothofagidites* spp. *fusca* group, and various bisaccate and trisaccate pollen of podocarp conifers. We note, however, that both of these groupings are wind-pollinated types and likely overrepresented in the flora, compared with Proteaceae and some of the other angiosperms that today produce relatively fewer pollen and are pollinated by insects and other means. Diversity of both the southern beech and conifer species is lower than observed in the latest Eocene.

The southern beech component is comprised mainly of cool to cold-climate *fusca* types *Nothofagidites lachlaniae* and *Nothofagidites* sp. cf. *Nothofagidites flemingii* (and related forms), as in NBP0602A-3C, and as observed in Oligocene and Neogene assemblages from the Ross Sea Cape Roberts Project (CRP) [*Raine*, 1998; *Askin and Raine*, 2000; *Raine and Askin*, 2001] and ANDRILL cores [*Taviani et al.*, 2008; *Warny et al.*, 2009]. Pollen of *Nothofagidites brassii* group (related to extant warm climate *Nothofagus* living in, for example, New Caledonia) are rare in the older NBP0602A-12A-2 samples, much the same as in NBP0602A-3C samples, and are likely reworked. Pollen of *Nothofagidites asperus* (*menziesii* group, related to extant mountain beech) are very rare (<1% in only a few samples), and again, it is not known if this type of beech grew in the Antarctic Peninsula in the Oligocene or if the pollen are reworked; this type exhibited only infrequent to rare occurrences in Seymour Island Eocene [*Askin*, 1997] and in latest Eocene NBP0602A-3C [*Warny and Askin*, this volume].

Podocarp conifer pollen include bisaccate species of *Podocarpidites*, *Phyllocladidites-Micro-alatidites* complex, and trisaccate *Podosporites-Trichotomosulcites* complex. One of the most common podocarp species is *Phyllocladidites* sp.a, with very small sacs, and many of the *Podosporites-Trichotomosulcites* complex pollen are morphotypes with barely developed vestigial sacs.

The presence of several to common specimens per sample (up to 4% of total in NBP0602A-12A-2, 202-204m) of simple *Proteacidites parvus-subscabratus* (Proteaceae) pollen types is significant, as it suggests that some Proteaceae survived through the Oligocene and possibly into the Neogene (see NBP0602A-5D discussion below) in the Antarctic Peninsula, confirming that the specimens observed in Ross Sea CRP [*Askin and Raine*, 2000; *Raine and Askin*, 2001] and ANDRILL [*Taviani et al.*, 2008; *Warny et al.*, 2009] assemblages may also be in place.

The terrestrial palynomorph assemblage includes a variety of other angiosperm pollen. These pollen include Liliaceae (*Liliacidites* spp., or other monocotyledon family with similar pollen), pollen of Chenopodiaceae/Caryophyllaceae (including *Chenopodipollis* sp. and pollen with fewer, more widely spaced pores more similar to Caryophyllaceae), Restionaceae, Poaceae, ?Ranunculaceae, and Asteraceae, and assorted other pollen of unknown affinities, which, along with moss spores including *Coptospora* spp. and *Stereisporites* spp., may have formed part of a tundra vegetation.

Numbers of penecontemporaneous terrestrial palynomorphs drop substantially in the upper two NBP0602A-12A-1 samples, both in total concentrations and in relation to marine palynomorphs, and the uppermost sample is almost all reworked.

Throughout these late Oligocene samples (and the Miocene samples discussed below), several different age and preservational groups are recognized among the reworked terrestrial specimens. Consistently occurring throughout much of the NBP0602A-12A and NBP0602A-5D sections, although in varying numbers, are well-preserved Late Cretaceous-Paleocene-Eocene pollen and spores. Sediments spanning these ages are well known from the nearby James Ross Basin, including Seymour Island, and imply northeastward-trending transport directions.

As noted for the Eocene NBP0602A-3C cores [*Warny and Askin*, this volume] and Eocene of Seymour Island [*Askin and Elliot*, 1982], scattered occurrences of well-preserved Permian-Triassic pollen and spores, such as the Permian spore *Granulatisporites trisinus* (Figure 3, microphotograph 11), are found in NBP0602A-12A and NBP0602A-5D. This suggests erosion and redeposition of no-longer exposed or preserved strata of that age that are relatively unmetamorphosed.

There are also intervals with common dark brownish black and black corroded palynomorphs, which are typically unidentifiable, except for pollen distinguishable as bisaccate and occasionally as possibly taeniate bisaccate (Figure 3, microphotograph 12). It is believed that this suite of reworked palynomorphs is of Permian and Triassic age, and likely derived from Permian-Triassic strata from the Antarctic Peninsula, that are somewhat more metamorphosed than the Permian-Triassic strata mentioned above. The Trinity Peninsula Series might be the source of this suite. For example, organic material extracted from Hope Bay is black and corroded and does not typically include identifiable palynomorphs. This suite of black corroded pollen and spores occurs in the lower two NBP0602A-12A samples (7% of the reworked component in both samples 11 and 12) and is especially notable in the upper two NBP0602A-12A samples (28% in sample 1 and 19% of reworked in sample 2).

2.3. Discussion on Ecology and Assemblage Composition

Several key observations were made regarding NBP0602A-12A. First, abundant dinoflagellate cysts of Albian to Maastrichtian age have been found, which indicate erosion of a succession of Cretaceous strata located "upstream" of the NBP0602A-12A drill site.

Second, the in situ phytoplankton specimens recovered are mainly composed of complexes of the acritarchs *Micrhystridium* spp. and *Leiosphaeridia* spp. This acanthomorph/sphaeromorph association is intriguing because it is one of the oldest fossil associations known on Earth, first recorded from Neoproterozoic rocks [*Chuanming et al.*, 2007]. These two groups are simple life forms, and they may well be opportunistic in otherwise abandoned environments. *Grey et al.* [2003] describe a major Ediacarian palynomorph species radiation (one species of leiospheres and 57 species of various acanthomorph acritarchs) that occurred around 580 Ma ago, following a major environmental disturbance, associated either with the postglacial phase of the "snowball Earth event" or the Acraman bolide impact event. In their study, these two groups of species are the main forms of life able to radiate after the major environmental disturbance.

The main third observation is that the penecontemporaneous terrestrial palynomorph assemblage is of lower diversity and much less abundant than the latest Eocene assemblage documented in Hole NBP0602A-3C. It is mainly composed of species of southern beech (*Nothofagidites* spp., *fusca* group) and podocarpaceous conifer pollen *Phyllocladidites* spp., with various angiosperms including Proteaceae, which likely formed a woodland to scrubby vegetation with a tundra association on more exposed sites. Overall, this flora suggests a much colder climate than indicated for the Eocene Hole NBP0602A-3C assemblages.

In summary, the presence in the marine realm of mostly leiospheres and acanthomorph acritarchs (both potential sea ice indicators and likely opportunistic in this deteriorating environment), the paucity of in situ dinoflagellate cysts, together with the terrestrial palynological signal, all indicate cold to polar marine environments with at least seasonal sea ice periodically at or near the depositional site. Clearly, the late Oligocene was a time of significant climate deterioration in the region, relative to the latest Eocene documented in Hole NBP0602A-3C.

3. THE MIOCENE SAMPLED BETWEEN 12.8 TO 11.8 MA (CORE NBP0602A-5D)

3.1. Material and Methods

Borehole NBP0602A-5D was drilled at 63°15.090′S latitude and 52°21.939′W longitude (Figure 7) and sampled the strata offshore and downdip of James Ross Island on the Joinville Plateau (Figure 8). The hole was drilled in 506 m of water depth (Shipboard Scientific Party, 2006) and is composed of a series of 13 drilled sections or cores (Figure 11), with a total length of ~30 m of cored section and ~13 m of recovered core. Figure 11 shows the intervals recovered and those of low to no recovery. As was the case for the Oligocene hole, continuous coring was made impossible by extremely harsh weather conditions and fast moving sea ice at the time of drilling. The Miocene core units are consecutively named core 13R_a (also referred to as NBP0602A-5D-13) from the base of the section to core 7R_a (also referred to as NBP0602A-5D-7) and core 6R_a at

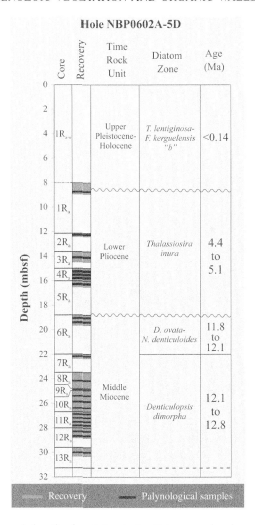

Figure 11. Palynological sample location in relation to the diatom zonal assignments and age interpretation for borehole NBP0602A-5D. Modified from Shipboard Scientific Party (2006).

the top. Based on diatom biostratigraphy [*Bohaty et al.*, this volume], units in cores $13R_a$ to $7R_a$ were assigned a maximum age of 12.8 Ma and a minimum age of 12.1 Ma based on the presence of the diatom *Denticulopsis dimorpha*. The overlying $6R_a$ section was assigned a maximum age of 12.1 Ma and a minimum age of 11.8 Ma based on the presence of two diatoms; *Denticulopsis ovata* and *Nitzschia denticuloides*. Fifteen samples were collected for palynological analysis. The location of palynological samples relative to the various core units recovered is indicated in Figure 11. As marked in Figure 11, samples were taken at regular intervals in each of the cored units recovered from $13R_a$ to $6R_a$.

Smith and Anderson [this volume] and *Wellner et al.* [this volume] both note that the sedimentology of the middle Miocene section ($13R_a$ to $6R_a$) represents a dramatic change in environment compared with older strata. They note that this unit contains a gray diamicton, representing proximal glacial marine sediments. The units sampled are composed of fine-grained muds alternating with a sandier greenish black silty mud lithology.

3.2. Palynological Results

3.2.1. General evaluation. Figure 12 presents a summary of palynological results obtained for the middle Miocene NBP0602A-5D SHALDRIL section (cores $13R_a$ to $7R_a$). Core $6R_a$, which is of younger middle Miocene age, is barren of penecontemporaneous palynomorphs. In general, the recovery is low but higher than from the NBP0602A-12A Oligocene section. The concentration of palynomorphs in Hole NBP0602A-5D is up to approximately 17,000 palyno- morphs (marine and terrestrial) per gram of dried sediment, in contrast to ~2000 in the Oligocene from NBP0602A-12A, and 120,000 palynomorphs per gram of dried sediments in the late Eocene recovered in drill hole NBP0602A-3C. Although higher than for the Oligocene section, the concentrations measured for the Miocene mostly reflect the presence of abundant reworked specimens and do not indicate a return to significantly better environmental conditions.

3.2.2. Dinoflagellate cysts and acritarchs. In this core, the *Leiosphaeridia* spp. complex is found only sporadically in a few samples. Acanthomorph acritarchs and small dinoflagellate cysts, belonging to the genera *Micrhystridium* and *Impletosphaeridium*, respectively, dominate most of the section. The latter genus is very similar in size and morphology to the acritarch *Micrhystridium*, except for the presence of an archeopyle (not always clearly visible) and tiny capitate process tips. Because *Micrhystridium* have a long stratigraphic range, it is important to examine whether these acanthomorph forms are in place or reworked. For instance, *Micrhystri- dium* are known to be extremely abundant in parts of the Late Cretaceous section [*Askin*, 1988; *Thorn et al.*, 2009], so reworking is a possibility. Careful examination of each specimen has allowed us to tabulate these acanthomorph acritarchs and dinoflagellate cysts as "in situ" or "reworked" (Figure 12). Those *Micrhystridium* specimens exhibiting a darker color (showing a golden hue) have been grouped as "reworked acanthomorph acritarchs." The acanthomorph acritarchs that are almost transparent, showing little maturation, and in excellent condition have been grouped as "in situ acanthomorphs" (Figure 2, microphotographs 4, 5, and 6). These associations are validated by the fact that peak abundances in reworked acanthomorph acritarchs tend to be associated with abundant reworked Cretaceous dinoflagellate cysts, or Cretaceous spores and pollen.

The cored intervals show changes in the reworked assemblage throughout the section, which indicates recycling input from more than one source. For example, sample 13 at the top of core $10R_a$ includes over 6000 dinocysts per gram of dried sediments, the majority of which are composed of very well preserved, though thermally mature (dark brown colored) Late Jurassic and Cretaceous species listed as "reworked Mesozoic" on Figure 12. These possibly indicate origin from a drop stone, although no sedimentological signal is seen in the geophysical proxies at this particular level [*Wellner et al.*, this volume] or a rather drastic shift in provenance for the short interval captured by this sample. Many of the diverse species that make up this assemblage are similar to species found on Byers Peninsula [*Askin*, 1981, 1983; *Duane*, 1996]. If this assemblage was deposited as a drop stone at Site NBP0602A-5D, it may have been eroded from Mesozoic outcrops located on the Antarctic Peninsula or western James Ross Basin. It is worth noting that the degree of maturation of most of these palynomorphs is more advanced (Figure 1, micro- photographs 1–6) than the maturation observed for other Cretaceous species (Figure 1, micro- photographs 7–13). Such an accelerated maturation can be linked to volcanic activity occurring in close proximity to the original depositional basin. On Byers Peninsula, the thermal maturation is clearly related to coeval volcanic activity. Volcanic activity is also known from the James Ross Basin and in the Antarctic Peninsula.

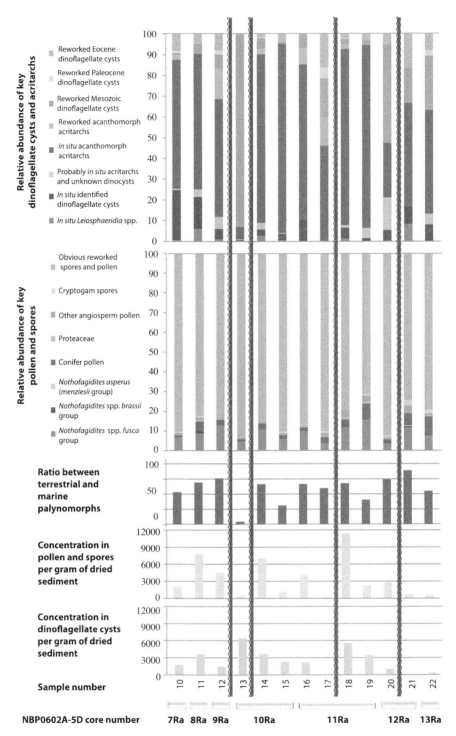

Figure 12. Relative abundance of key palynomorphs, concentration in palynomorphs per gram of dry sediment, and ratio between terrestrial (gray bars) and marine palynomorphs in core NBP0602A-5D. The red lines separate zones discussed in the text where palynological changes are observed.

Cores 9R$_a$, 8R$_a$, and 7R$_a$ (samples 12–10), although also dominated by acanthomorph acritarchs, show a marked increase in dinoflagellate species that are believed to be in place. These include specimens of *Spiniferites ramosus*, *Operculodinium centrocarpum*, and a species we include as *Pyxidinopsis braboi*. The wall structure of the latter species is very similar to that of *Filisphaera filifera*, with densely distributed membranous septa arranged in a reticulum and a precincular archeopyle, but all our specimens lack the apical lobe typical of *F. filifera*. *O. centrocarpum* and *P. braboi* are identical to the species found in abundance at 15.7 Ma during an episode we interpreted as indicative of a warmer interval in the McMurdo Sound ANDRILL SMS cores [*Warny et al.*, 2009]. *O. centrocarpum* was also found in Miocene sections in Bruce Bank [*Mao and Mohr*, 1995] and in other Miocene sections of the Antarctic Peninsula [*Troedson and Riding*, 2002]. Presence of these dinoflagellate cysts indicate somewhat improved conditions in the marine realm, compared with Miocene samples below core 9R$_a$, and/or input of nutrients in the Weddell Sea that would allow for dinoflagellate blooms.

The majority of reworked specimens found in the NBP0602A-5D drill hole are of Jurassic/ Lower Cretaceous, Late Cretaceous, and Eocene age. Specimens grouped as "Reworked Mesozoic dinoflagellate cysts" are the most abundant. Some specimens were quite damaged, in which case, identification was difficult. Mesozoic species observed include ?*Rigaudella aemula* (?Early Cretaceous), *Prolixosphaeridinum inequiornatum* (Aptian), *Batiolodinium radiculatum* (Aptian), *Muderongia tetracantha* (Neocomian-lower Albian), *D. cristatum* (Aptian-Cenomanian), *C. compactum* (Albian-Santonian), *O.* cf. *pulcherrimum* (?Albian), *O. operculata* (Aptian-Senonian), *O. porifera* (Senonian), *C. victoriensis* (Turonian-Santonian), *O. spinosa* (Maastrichtian), *Isabelidinium cretaceum* (Santonian-Maastrichtian), *M. seymourensis* (Maastrichtian), *Manumiella seelandica* (late Maastrichtian-earliest Danian), *M. druggii* (late Maastrichtian-earliest Danian), ?*Hystrichogonaulax serrata*, various species that are most likely associated to *Cribroperidinium* spp. (including *C. longicorne*, a Late Jurassic species), *Glaphyrocysta* spp., and *Apteodinium* spp. (including *A. punctatum*), and the acritarch *V. reductum* (Cretaceous). *S.* cf. *lanternum* may represent Paleocene age. The specimens grouped as reworked Eocene species are *O. askiniae* (Eocene), *V. apertura* (Paleocene-Oligocene), *E. partridgei* (middle Eocene to early Oligocene), *S. macmurdoense* (early Tertiary), *Turbiosphaera filosa* (Paleocene to Oligocene), *D. antarctica* (Eocene), and *Stoveracysta ornata* (late Eocene).

3.2.3. Pollen and spores. The middle Miocene NBP0602A-5D assemblages show a substantial decrease in numbers and diversity of presumed penecontemporaneous specimens, compared with the Oligocene samples. Instead, the assemblages are overwhelmingly reworked. Dilution by rapid sedimentation rate may have caused the very low concentration in some samples. Alternatively, these may have been times when vegetation was sparse or essentially lacking, only existing in small protected pockets on the Antarctic Peninsula.

The terrestrial assemblages, like those in the NBP0602A-12A cores, are dominated by *Nothofagidites* spp. of the *fusca* group, especially *N. lachlaniae*. Podocarp conifer pollen is of secondary importance and become less frequent in the younger samples, with *Phyllocladidites* sp.a being the most common species. Simple Proteaceae pollen are rare (1% or less, or absent, but up to 2% in sample 14). All other terrestrial taxa together are relatively common (total up to 4%) in the lower five samples, then become very rare (1% or less or absent). They include pollen and spores of probable tundra plants such as Poaceae (grasses), Chenopodiaceae, Caryophyllaceae (including specimens of *Colobanthus*-type), Asteraceae, and moss spores.

We note that a specimen in the youngest sample (10, NBP0602A-5D-7, 27-29 m, Figure 6, number 8) is similar to pollen of extant *Colobanthus quitensis*. This is one of only two species of higher plants that survive in protected localities on the northern Antarctic Peninsula today, as does one grass, *Deschampsia antarctica*.

As noted in the Oligocene discussion above, there are several suites of reworked terrestrial palynomorphs, including the Late Cretaceous and Eocene suites, the well-preserved Permian-Triassic suite, and the thermally mature presumed Permian-Triassic suite. Scattered rare occurrences of the latter group occur in NBP0602A-5D cores, with increased numbers of these black, corroded, thermally altered palynomorphs being recorded in samples 13 and 17 (7% and 10% of the reworked component, respectively) and in sample 21 (4%).

The suite of Early Cretaceous palynomorphs are moderately thermally altered and are of orange to orange-brown to brown in color and are composed primarily of spores with some pollen, typical of Early Cretaceous assemblages. They include the cryptogam spores *Cyatheacidites annulatus*, *Ciactricosisporites* spp., *Contignisporites* spp., *Murospora florida*, a variety of apiculate and verrucate spores, plus saccate conifer pollen, and the distinctive conifer pollen *Classopollis classoides*. This suite is similar in species composition and thermal maturation to those recovered from Byers Peninsula, Livingston Island, and is likely derived from similar aged sediments on the flanks of the Antarctic Peninsula-western James Ross Island. These Early Cretaceous pollen and spores occur sporadically throughout the NBP0602A-5D cores, but are especially common in sample 13, associated with the abundant reworked Early Cretaceous dinoflagellate cysts discussed above.

3.3. Discussion on Ecology and Assemblage Composition

SHALDRIL Hole NBP0602A-5D recovered sediments that have been dated as middle Miocene to possibly late Miocene (~12.8 to ~11.8 Ma) by diatom stratigraphy [*Bohaty et al.*, this volume]. The abundance of in situ acanthomorph acritarchs, mostly *Micrhystridium* spp., characterizes this Miocene interval. Little is known about the ecology of these acanthomorph forms, but from what we see in their distribution, we argue that their behavior in the ocean is similar to that of weeds on land. They seem to be opportunistic species that take over environments under stress when the local flora can no longer subsist and predation pressure decreases, allowing them to thrive when everything else is struggling. But with so little information available on their ecology, it is difficult to understand the environmental significance of the high abundance observed. *Warny et al.* [2006, 2007] reviewed whether the morphology of a dinocyst can be linked to environmental sea-surface parameters by analyzing the distribution of some extant species with similar morphologies. Although the form discussed herein is an acritarch, a similar ecomorphological approach could be used. Extant dinoflagellate cysts of *Echinidinium* spp, *Islandinium cezare*, *Islandinium minutum*, and *Pentapharsodinium dalei* all have a similar overall morphology to *Micrhystridium* sp. and other forms such as the dinoflagellate *Impletosphaeridum* sp. Data on the modern ecological distribution of the four modern species showed that these species can all be present (some in high relative abundance) when sea ice is present. *Echinidinium* spp. can make up to 8% of the assemblage in areas with 8 to 11 months of sea ice cover; *I. cezare*, *Echinidinium karaense*, and *Islandinium minutum* are often predominant in surface waters covered with sea ice between 6 and 12 months of the year. *P. dalei* has a broader range of distribution and can be found in waters covered with sea ice from 1 to 12 months per year (Figure 13). If this morphologic-similarity hypothesis is correct, and if the high abundance in acanthomorph forms is indeed indicative of a specific set of paleoenvironmental conditions, then it most likely means that these species, like their modern counterparts, can thrive in environments associated with ephemeral sea ice development. This observation is supported by sedimentological data. Sediments of the middle part of the section in Hole NBP0602A-5D (~12.1 to ~11.8 Ma)

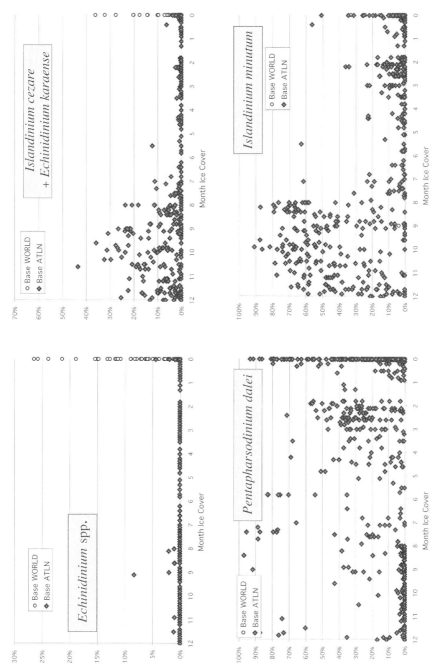

Figure 13. Modern relative abundance of selected dinoflagellate cyst species against months of sea ice cover, From *Warny et al.* [2007].

include soft pebbly gray diamicton, with no signs of burrowing, and they are interpreted as proximal glaciomarine deposits [*Wellner et al.*, this volume].

A single glacial source is suggested for core NBP0602A-5D-10 where a rich Cretaceous assemblage has been found, possibly brought as a dinocyst-rich drop stone. This single level high abundance in Cretaceous dinoflagellate cysts and spores could indicate a significantly different provenance because of the marked difference between the reworked specimens at this level as opposed to the type of reworked specimens found at other levels. It could also indicate a more severe glacial advance and erosion of Cretaceous strata at that single level.

Conditions at the top of the core seem to be more favorable with the presence of a few dinoflagellate species that were found in great abundance during the Mid-Miocene Climatic Optimum in the ANDRILL core collected at McMurdo Sound [*Warny et al.*, 2009]. They are not abundant in the present study, with just a few hundred cysts per gram of dried sediment, but their presence might indicate that the ocean was not covered by sea ice all year long and that sufficient nutrients were present in the surface waters.

The pollen and spores recovered from this Miocene section are mostly reworked specimens with sparse penecontemporaneous specimens of *Nothofagidites* spp. (*fusca* group) and podocarpaceous conifer pollen (the most common of which is *Phyllocladidites* sp.a), with presumed tundra taxa (identified as such in Ross Sea-Dry Valleys Neogene assemblages by *Askin and Raine*, 2000, *Lewis et al.*, 2008, and *Warny et al.*, 2009) such as Chenopodiaceae, Caryophyllaceae (*Colobanthus*-type), and Poaceae (grasses). As mentioned above, the latter two taxa are plants that grow today on the Antarctic Peninsula in addition to various moss species. It is likely that woody taxa (such as the southern beech and conifers) mostly had a low shrubby to ground-hugging habit in these more extreme conditions. The Miocene vegetation during the interval sampled was likely composed of mossy tundra, with low shrubby trees/bushes in more protected areas, similar to that described for the Neogene of the Ross Sea area.

These sparse marine and terrestrial palynomorphs indicate that the middle Miocene section sampled by SHALDRIL Hole NBP0602A-5D may represent the closing stages in the climatic deterioration of the region. These results are consistent with the seismic unconformity located off the northwestern Peninsula and interpreted by *Smith and Anderson* [this volume] as the record of the first major offshore expansion of the APIS. *Anderson et al.* [2011] argue that this unconformity is part of the phased northward expansion of the APIS on the eastern side of the Peninsula, which began in the late middle Miocene to late Miocene in the southern part of James Ross Basin and expanded onto the northern part of the basin by latest Miocene to early Pliocene time.

4. PLIOCENE SAMPLED BETWEEN 5.1 AND 3.8 MA
(CORES NBP0602A-5D, 6C, AND 6D)

4.1. Material and Methods

The Pliocene interval analyzed for this study was sampled in two separate sections. The first section is from the previously described borehole NBP0602A-5D and comprises the top five cored sections (5Ra from the base of the section to core 1Ra at the top). Based on the presence of the diatom *Thalassiosira inura* [*Bohaty et al.*, this volume], these units are assigned a maximum lower Pliocene age of 5.1 Ma and a minimum age of 4.4 Ma. Seven samples were collected for palynological analysis from this lower Pliocene unit. The location of palynological samples relative to the various core units recovered is indicated on Figure 11. As marked in Figure 11, samples were taken at regular intervals in each of the cored units recovered from NBP0602A-5D-5 to NBP0602A-5D-1.

The lithologies of the lower Pliocene units include mud to muddy fine sands, with a thick unit of silty diatomaceous mud [*Wellner et al.*, this volume]. The latter sediment type has previously proved to be excellent for palynomorph recovery in Antarctica [*Warny et al.*, 2009].

The second section analyzed was recovered from boreholes NBP0602A-6C and 6D, drilled immediately south of NBP0602A-5D in the Joinville Plateau (Figures 7 and 8). Borehole NBP0602A-6C was drilled at 63°20.268′S latitude and 52°22.032′W longitude. The hole was drilled through 532 m of water depth (Shipboard Scientific Party, 2006). This hole is composed of a series of nine drilled sections or cores, seven of which recovered a total of 5.9 m of sediment, and two had no recovery (Figure 14). The core units are consecutively named core 9Ra from the base of the section to core 1Ra at the top. Eighteen samples were collected for palynological analysis from this lower Pliocene unit: 10 from NBP0602A-6C and 8 from NBP0602A-6D.

Borehole NBP0602A-6D was drilled at 63°19.746′S latitude and 52°22.040′W longitude (Figures 7 and 8), through 528 m of water depth (Shipboard Scientific Party, 2006). This hole is composed of a series of three drilled sections or cores and has a total length of 2.27 m of core recovered. Figure 14 shows intervals recovered and those of low to no recovery in between. The core units are consecutively named core 3Ra from the base of the section to core 1Ra at the top.

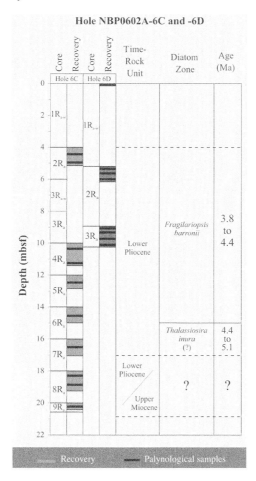

Figure 14. Palynological sample location in relation to the diatom zonal assignments and age interpretation for borehole NBP0602A-6C and 6D. Modified from Shipboard Scientific Party (2006).

Based on diatom biostratigraphy [*Bohaty et al.*, this volume], units in cores NBP0602A-6C-7 to NBP0602A-6C-2 and in cores NBP0602A-6C-3 to NBP0602A-6C-2 were assigned a maximum age of 5.1 Ma and a minimum age of 3.8 Ma based on the *T. inura* and the *Fragilariopsis barronii* diatom zones. Borehole NBP0602A-6C is composed of pebbly to muddy sand, while borehole NBP0602A-6D is a sandy mud.

4.2. Palynological Results

The seven samples from the upper NBP0602A-5D series, and the 18 samples from the NBP0602A-6C and NBP0602A-6D series, all dated early Pliocene, were processed but were essentially barren of penecontemporaneous palynomorphs. Obvious reworked specimens of different ages were found in varying but low numbers in most samples. Of importance in the top sample of core NBP0602A-6C-4 (Figure 14) is the occurrence of several specimens of Chenopodiaceae and an Asteraceae pollen, which may be in place and would hence represent vestiges of a surviving vegetation. But without higher concentrations, it is impossible to determine whether these are in place or reworked from the Miocene.

4.3. Discussion on Ecology and Assemblage Composition

Holes NBP0602A-6C and NBP0602A-6D and top of Hole NBP0602A-5D sampled early Pliocene deposits above a major glacial unconformity (Figure 8) [*Anderson et al.*, 2011]. The paucity of palynomorphs in the Pliocene sections analyzed can be the result of four factors. First, the samples could appear barren because of extremely high sedimentation rate that would dilute the assemblage; second, palynomorphs could have been winnowed out of the sandier sediment; third, it could be due to poor preservation; or fourth, it could simply denote a lack of vegetation on land and absence of organic microphytoplankton in the ocean because of long-term ice cover.

Because of the high latitude of this site, it is unlikely that sedimentation rates were extremely high at the time of deposition during glacial times. The second possibility is that the lack of palynomorphs results from sedimentological factors. The majority of the section is composed of greenish-gray sandy units with minor clay and abundant ice-rafted clasts of variable lithology. *Anderson et al.* [2011] interpreted this type of sediment to represent a current-winnowed deposit with iceberg-rafted material from distant sources, similar to the modern sediments of the area. It would hence be possible that the palynomorphs, most of which are under 100 μm in diameter, were washed out of the sandier lithology, but the diatomaceous muds that were also sampled within these boreholes are also barren of palynomorphs. Furthermore, the Miocene section recovered in Hole NBP0602A-12A consisted of sandy mud, and palynomorphs were recovered from that site, despite the sandy lithology. The third possibility is that the palynomorphs were not preserved. We do not have any means to test this, but the fact that some reworked (although rare) palynomorphs were found in these samples would tend to contradict this. Hence, we favor the hypothesis that the lack of fossil evidence is the result of the demise of the vegetation on nearby land and in the ocean. This conclusion agrees with the work of *Anderson et al.* [2011] that, based on geophysical evidence, suggests that by the end of the Miocene, ice extended to the north of the Antarctic Peninsula. The lack of palynomorphs would thus be explained by the fact that most of the area was ice covered and that the vegetation and phytoplankton were essentially absent or dormant and unable to reproduce. Other seismic data acquired in the Antarctic Peninsula area [*Bart and Anderson*, 2000; *Bart et al.*, 2005, 2007] reveal that the APIS was a dynamic ice sheet, with a minimum of 10 expansions and contractions of the APIS to the Joinville Plateau during the Pliocene-Pleistocene. Based on the

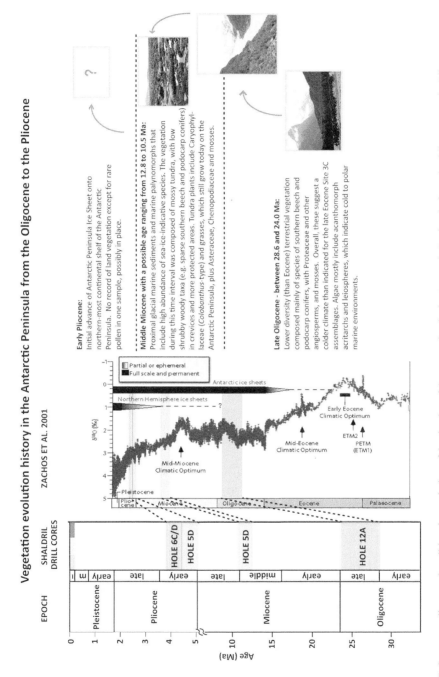

Figure 15. Summary diagram describing the evolution of the demise of the vegetation in the Antarctic Peninsula from the Oligocene to the Pliocene. Data curve is from *Zachos et al.* [2001]. Photographs by R. Askin (top and bottom) and H. Morgans (middle).

above evidence, it is clear that environmental conditions at the time of deposition were mostly unsuitable for land plants and dinoflagellate algal blooms in the Joinville Plateau area.

Nevertheless, the lack of palynomorphs in these sequences does not preclude the existence of vegetation pockets in some isolated parts of the Peninsula, which might explain the recovered specimens of possible penecontemporaneous pollen of opportunistic "weedy" plants (Chenopodiaceae and Asteraceae). These insect-pollinated plants are known to produce only low concentrations of pollen, further hampering the chances to firmly prove the existence of remnant pockets of Pliocene vegetation.

5. CONCLUSIONS

Three key stratigraphic intervals were sampled on the Joinville Plateau. They provide some glimpses into the demise of the vegetation from the late Oligocene, the middle Miocene, to the early Pliocene (Figure 15).

In both the Oligocene and the Miocene, small sea ice-tolerant, endemic, opportunistic species of acritarchs form the bulk of the organic phytoplankton present in the Joinville Plateau, along with rare dinoflagellate cysts that most likely adapted to the cold, polar conditions. During this time, the terrestrial palynomorph assemblages indicate sparse to very limited vegetation on the adjacent Antarctic Peninsula landmass. Vegetation during deposition of the late Oligocene cores, like the latest Eocene vegetation [*Warny and Askin*, this volume], was dominated by southern beech, with secondary podocarp conifers. By the time the middle Miocene section cores were deposited, ice covered much of the Antarctic Peninsula [*Anderson et al.*, 2011], and terrestrial plant life was evidently reduced to pockets of vegetation, most likely tundra that included stunted woody plants such as southern beech and podocarp conifers, and a few other herbaceous angiosperms and mosses. Tundra vegetation may have survived longer here than in the Victoria Land-Ross Sea region, where *Marchant et al.* [2009] found cold desert landforms (polygons) with ash-filled wedges providing a date of 13.85 Ma, indicating that by that time, a permanent polar desert existed in the region and that the Dry Valley tundra disappeared from the region. Our record shows that tundra vegetation existed on the Antarctic Peninsula up to at least 12.8 Ma, making this region possibly the last refugium for vegetation in Antarctica, about 1 million years later than the last tundra record [*Lewis et al.*, 2008] found in the Ross Sea region.

Today, *Colobanthus* and grass grow in sheltered areas on the northern Antarctic Peninsula. Their pollen occurs in Miocene sediments but thus far has not been found in the Pliocene. Further work is needed to determine if these plants might have been extinguished from the Antarctic Peninsula and then recolonized (possibly from the South Shetland Islands) or if small pockets of vegetation survived unscathed until today.

Acknowledgments. Thanks are extended to the National Science Foundation for funding this research (grant from NSF Office of Polar Programs 0636747) and to John Anderson of Rice University and Julia Wellner of the University of Houston for providing the samples necessary for the completion of this study. Thanks are also extended to the Antarctic Research Facility staff for giving us access to the cores and assisting us with sampling needs. We thank the two reviewers, David Jarzen and Vanessa Bowman, who provided a careful review of this manuscript.

REFERENCES

Anderson, J. B., et al. (2011), Progressive Cenozoic cooling and the demise of Antarctica's last refugium, *Proc. Natl. Acad. Sci. U. S. A.*, *108*(28), 11,356–11,360, doi:10.1073/pnas.1014885108.

Askin, R. A. (1981), Jurassic-Cretaceous palynology of Byers Peninsula, Livingston Island, Antarctica, *Antarct. J. U. S.*, *16*(5), 11–13.

Askin, R. A. (1983), Tithonian (uppermost Jurassic)-Barremian (Lower Cretaceous) spores, pollen and microplankton from the South Shetland Islands, Antarctica, in *Antarctic Earth Science: Australian Academy of Science*, edited by R. L. Oliver, P. R. James, and J. B. Jago, pp. 295–297, Cambridge Univ. Press, Cambridge, U. K.

Askin, R. A. (1988), Campanian to Paleocene palynological succession of Seymour and adjacent islands, northeastern Antarctic Peninsula, *Mem. Geol. Soc. Am.*, *169*, 131–154.

Askin, R. A. (1997), Eocene - ?earliest Oligocene terrestrial palynology of Seymour Island, Antarctica, in *The Antarctic Region: Geological Evolution and Processes*, edited by C. A. Ricci, pp. 993–996, Terra Antart. Publ., Siena, Italy.

Askin, R. A., and D. H. Elliot (1982), Geologic implications of recycled Permian and Triassic palynomorphs in Tertiary rocks of Seymour Island, Antarctic Peninsula, *Geology*, *10*, 547–551.

Askin, R. A., and J. I. Raine (2000), Oligocene and early Miocene terrestrial palynology of the Cape Roberts drillhole CRP-2/2A, Victoria Land Basin, Antarctica, *Terra Antart.*, *7*(4), 493–501.

Bart, P. J., and J. B. Anderson (1995), Seismic record of the glacial events affecting the Pacific margin of the northwestern Antarctic Peninsula, in *Geology and Seismic Stratigraphy of the Antarctic Margin*, *Antarct. Res. Ser.*, vol. 68, edited by A. K. Cooper et al., pp. 75–95, AGU, Washington, D. C.

Bart, P. J., and J. B. Anderson (2000), Relative temporal stability of the Antarctic ice sheets during the late Neogene based on the minimum frequency of outer shelf grounding events, *Earth Planet. Sci. Lett.*, *182*(3–4), 259–272.

Bart, P. J., D. Egan, and S. Warny (2005), Direct constraints on Antarctic Peninsula Ice Sheet grounding events between 5.12 and 7.94 Ma, *J. Geophys. Res.*, *110*, F04008, doi:10.1029/2004JF000254.

Bart, P. J., C. D. Hillenbrand, W. Ehrmann, M. Iwai, D. Winter, and S. Warny (2007), Are Antarctic Peninsula Ice Sheet grounding events manifest in sedimentary cycles on the adjacent continental rise?, *Mar. Geol.*, *236*(1–2), 1–13.

Birkenmajer, K. (1996), Tertiary glacial/interglacial paleoenvironments and sea-level changes, King George Island, West Antarctica, An overview, *Bull. Pol. Acad. Sci. Earth Sci.*, *44*, 157–181.

Bohaty, S. M., D. K. Kulhanek, S. W. Wise Jr., K. Jemison, S. Warny, and C. Sjunneskog (2011), Age assessment of Eocene–Pliocene drill cores recovered during the SHALDRIL II expedition, Antarctic Peninsula, in *Tectonic, Climatic, and Cryospheric Evolution of the Antarctic Peninsula*, doi:10.1029/2010SP001049, this volume.

Chuanming, Z., X. Guwei, K. McFadden, X. Shuhai, and Y. Xunlai (2007), The diversification and extinction of Doushantuo-Pertatataka acritarchs in South China: Causes and biostratigraphic significance, *Geol. J.*, *42*, 229–262.

Dingle, R. V., and M. Lavelle (1998), Antarctic Peninsular cryosphere: Early Oligocene (c. 30 Ma) initiation and a revised glacial chronology, *J. Geol. Soc.*, *155*, 433–437.

Duane, A. (1996), Palynology of the Byers Group (Late Jurassic Early Cretaceous) of Livingston and Snow Islands, Antarctic Peninsula: Its biostratigraphical and palaeoenvironmental significance, *Rev. Palaeobot. Palynol.*, *91*(1–4), 241–281.

Grey, K., M. R. Walter, and C. R. Calver (2003), Neoproterozoic biotic diversification: Snowball Earth or aftermath of the Acraman impact?, *Geology*, *31*(5), 459–462.

Lewis, A. R., et al. (2008), Mid-Miocene cooling and the extinction of tundra in continental Antarctica, *Proc. Natl. Acad. Sci. U. S. A.*, *105*(31), 10,676–10,680.

Mao, S., and B. A. R. Mohr (1995), Middle Eocene dinocysts from Bruce Bank (Scotia Sea, Antarctica) and their paleoenvironmental and paleogeographic implications, *Rev. Palaeobot. Palynol.*, *86*, 235–263.

Marchant, D., A. Lewis, A. Ashworth, D. Kowalewski, K. Swanger, J. Willenbring, and S. Mackay (2009), The glacial and climate record of the Dry Valleys, Southern Victoria Land, paper presented at First Antarctic Climate Evolution Symposium, Antarct. Clim. Evol., Granada, Spain.

Mudie, P. J. (1992), Circum-arctic Quaternary and Neogene marine palynofloras: Paleoecology and statistical analysis, in *Neogene and Quaternary Dinoflagellate Cysts and Acritarchs*, edited by M. J. Head and J. H. Wrenn, pp. 347–390, Am. Assoc. Stratigr. Palynol. Found., Dallas, Tex.

Raine, J. I. (1998), Terrestrial palynomorphs from Cape Roberts Project drillhole CRP-1, Ross Sea, Antarctica, *Terra Antart.*, *5*(3), 539–548.

Raine, J. I., and R. A. Askin (2001), Oligocene and early Miocene terrestrial palynology of the Cape Roberts Drillhole CRP-2/2A, Victoria Land Basin, Antarctica, *Terra Antart.*, *7*, 389–400.

Smith, R. T., and J. B. Anderson (2011), Seismic stratigraphy of the Joinville Plateau: Implications for regional climate evolution, in *Tectonic, Climatic, and Cryospheric Evolution of the Antarctic Peninsula*, doi:10.1029/2010SP000980, this volume.

Taviani, M., et al. (2008), Palaeontologic Characterization of the AND-2A Core, ANDRILL Southern McMurdo Sound Project, Antarctica, in *Studies From the ANDRILL, Southern McMurdo Sound Project, Antarctica*, edited by D. M. Harwood et al., *Terra Antart.*, *15*, 113–144.

Thorn, V. C., J. B. Riding, and J. E. Francis (2009), The Late Cretaceous dinoflagellate cyst *Manumiella*—Biostratigraphy, systematics, and palaeoecological signals in Antarctica, *Rev. Palaeobot. Palynol.*, *156*(3–4), 436–448.

Troedson, A. L., and J. B. Riding (2002), Upper Oligocene to lowermost Miocene strata of King George Island, South Shetland Islands, Antarctica: Stratigraphy, facies analysis, and implications for the glacial history of the Peninsula, *J. Sediment. Res.*, *72*(4), 510–523.

Warny, S., and R. Askin (2011), Vegetation and organic-walled phytoplankton at the end of the Antarctic greenhouse world: Latest Eocene cooling events, in *Tectonic, Climatic, and Cryospheric Evolution of the Antarctic Peninsula*, doi:10.1029/2010SP000965, this volume.

Warny, S., J. H. Wrenn, P. J. Bart, and R. A. Askin (2006), Palynological analysis of the NBP03-01A transect in Northern Basin, western Ross Sea, Antarctica: A late Pliocene record, *Palynology*, *30*, 151–182.

Warny, S., J. B. Anderson, L. Londeix, and P. J. Bart (2007), Analysis of the dinoflagellate cyst genus *Impletosphaeridium* as a marker of sea-ice conditions off Seymour Island: An ecomorphological approach, in Antarctica: A Keystone in a Changing World—Online Proceedings of the 10th International Symposium on Antarctic Earth Sciences, edited by A. K. Cooper, C. R. Raymond, and 10th ISAES Editorial Team, *U.S. Geol. Surv. Open File Rep., 2007-1047*, 4 pp., doi:10.3133/of2007-1047.srp079. [Available at http://pubs.usgs.gov/of/2007/1047/srp/srp079/]

Warny, S., R. A. Askin, M. Hannah, B. Mohr, I. Raine, D. M. Harwood, F. Florindo, and the SMS Science Team (2009), Palynomorphs from a sediment core reveal a sudden remarkably warm Antarctica during the middle Miocene, *Geology*, *37*(10), 955–958, doi:10.1130/G30139A.1.

Wellner, J. S., J. B. Anderson, W. Ehrmann, F. M. Weaver, A. Kirshner, D. Livsey, and A. Simms (2011), History of an evolving ice sheet as recorded in SHALDRIL cores from the northwestern Weddell Sea, Antarctica, in *Tectonic, Climatic, and Cryospheric Evolution of the Antarctic Peninsula*, doi:10.1029/2010SP001047, this volume.

Wrenn, J. H., and G. F. Hart (1988), Paleogene dinoflagellate cyst biostratigraphy of Seymour Island, Antarctica, *Mem. Geol. Soc. Am.*, *169*, 321–447.

Zachos, J. C., M. Pagani, L. Sloan, E. Thomas, and K. Billups (2001), Trends, rhythms, and aberrations in global climate 65 Ma to present, *Science*, *292*, 686–693, doi:10.1126/science.1059412.

R. Askin, 1930 Bunkhouse Dr., Jackson, WY 83001, USA.

S. Warny, Department of Geology and Geophysics and Museum of Natural Science, Louisiana State University, Baton Rouge, LA 70803, USA. (swarny@lsu.edu)

Vegetation and Organic-Walled Phytoplankton at the End of the Antarctic Greenhouse World: Latest Eocene Cooling Events

Sophie Warny

*Department of Geology and Geophysics and Museum of Natural Science, Louisiana State University
Baton Rouge, Louisiana, USA*

Rosemary Askin

Jackson, Wyoming, USA

The NBP0602A-3C SHALDRIL section collected in the Weddell Sea recovered a key stratigraphic interval that captured the response of plants and organic-walled phytoplankton during the first marked increase in $\delta^{18}O$ followed by the large reduction in atmospheric CO_2 in the late Eocene around 36 Ma. Well-preserved palynomorphs recovered from in situ shelf sediments provide evidence of a cooling event followed by a marked sea level drop around 36 Ma. Terrestrial palynomorphs indicate that at the time of deposition, southern beech-dominated and conifer forest vegetation was abundant but with lower diversity and signifying colder climates than for most of the La Meseta Formation on Seymour Island. The marine palynomorph assemblage is dominated by *Vozzhennikovia apertura*. This low-diversity, high-dominance dinoflagellate cyst assemblage is also a sign of deteriorating conditions. Particularly notable is the marked increase in the uppermost Eocene samples of reworked dinoflagellates and acritarchs of Cretaceous age. This suggests significant erosion and redeposition of nearby Campanian-Maastrichtian sections during a marked drop in sea level. Based on the biostratigraphy and a single isotopic date, it is likely that the cooling and subsequent lowering of sea level can be correlated to the brief spike in $\delta^{18}O$-enriched values shown by the Zachos et al. (2001) curve in the Priabonian. According to Zachos et al. (2008), this event occurs at a time when lowest carbon dioxide atmospheric concentrations were between 600 and 980 ppmv, giving us some perspectives as to what could be expected when the current CO_2 atmospheric concentration is at least doubled.

Tectonic, Climatic, and Cryospheric Evolution of the Antarctic Peninsula
Special Publication 063
10.1029/2010SP000965

1. INTRODUCTION

Palynological investigations in Antarctica have been one of the primary means of elucidating shifts in past Antarctic climate [*Warny et al.*, 2006, 2009; *Thorn et al.*, 2009; *Anderson et al.*, 2011]. One such climatic shift of interest is when and how the climate evolved from greenhouse to icehouse conditions around the Eocene-Oligocene boundary, approximately 33.7 Myr ago. To refine our understanding of this important transition, one of the goals of the SHALDRIL II campaign was to drill through a section believed to be composed of in situ sediment that recorded changing climatic conditions across the Eocene-Oligocene boundary, and Borehole NBP0602A-3C was acquired. This chapter focuses on the palynological results of SHALDRIL II Borehole NBP0602A-3C, northern Antarctic Peninsula. We summarize and discuss recovered terrestrial fossil palynomorphs (pollen and spores), which help reconstruct past vegetation and thus paleoclimate, and marine palynomorphs (mostly dinoflagellate cysts and acritarchs), which are useful proxies for reconstructing sea-surface conditions such as sea-surface salinities, sea-surface temperature, and sea ice cover. Although palynomorph recovery is highly variable in many Eocene to recent Antarctic marginal strata, assemblages recovered are extremely valuable for detailed biostratigraphic and paleoenvironmental reconstruction. For instance, *Mao and Mohr* [1995], thanks to exceptionally rich dinocyst assemblages from Bruce Bank, were able to estimate winter surface water paleotemperatures at about 5°C to 10°C and summer temperatures reaching more than 14°C for the Neogene climatic optimum. *Warny et al.* [2009] highlighted the existence of a short and sudden warm climatic event 15.7 Myr ago, based on marine and terrestrial palynomorphs recovered from ANDRILL-2/2A in the Ross Sea. In this study, palynomorph assemblages provided further quantitative data of what the environment in Antarctica was like during the Mid-Miocene Climatic Optimum. They indicated that annual sea-surface temperatures ranged from 0°C to 11.5°C, land temperatures reached 10°C (January mean), and an increase in meltwater (and possibly rainfall) producing ponds and lakes adjacent to the Ross Sea developed during a short period of sea ice reduction.

In applying similar techniques to the SHALDRIL NBP0602A-3C section, a primary focus of the current study was to determine if major environmental changes were manifested first by changes of climate on land (this would be visible in terrestrial palynomorph changes), or did the changes occur first in the marine realm (as would be visible in changes in dinoflagellate cyst and acritarch assemblages), or were the changes occurring concurrently on land and in the ocean? One fundamental question we posed ourselves was "Did sea-surface cooling precede climate (land) cooling?" Better constraining the timing of these environmental changes both in the terrestrial and marine realms would help us better understand the source of these changes. The assumption here is that if opening of ocean passages were the main factor driving Antarctic climate change [*Kennett*, 1977], then thermal isolation would have affected dinocysts first since these organisms are directly influenced by ocean currents. Conversely, if decreased atmospheric CO_2 was the main control on the cryosphere's development [*DeConto and Pollard*, 2003], then initial cooling should have affected land plants first. If both marine and terrestrial organisms are affected at the same time, within limits of resolution, then one might conclude that the end of the greenhouse world was a consequence of both gateway closure and lower concentration in atmospheric carbon dioxide, with both factors possibly interrelated.

2. MATERIAL AND METHODS

Borehole NBP0602A-3C's location was selected specifically because seismic profiles show a series of basinward-tilted sedimentary deposits truncated at/near the seafloor. These strata were estimated to range from upper Eocene to upper Neogene in age [*Anderson*, 1999; *Anderson et al.*,

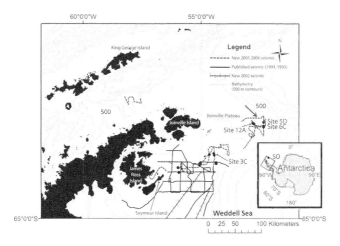

Figure 1. Location of SHALDRIL NBP0602A-3C drill hole off James Ross Island, Weddell Sea, Antarctica. From *Anderson et al.* [2011].

2006; *Anderson et al.*, 2011]. Based on these seismic data, Anderson and colleagues concluded that the Eocene section should be accessible at relatively shallow depth as recent glacial erosion exposed this stratigraphic section below a thin (less than one meter) glacial unit deposited at the seafloor. This section was thus a prime target for a SHALDRIL-type drilling campaign.

Borehole NBP0602A-3C was drilled at 63°50.861′S latitude and 54°39.207′W longitude and sampled the strata immediately offshore and downdip of James Ross Island in the Weddell Sea (Figure 1). This hole is composed of a series of seven drilled sections or cores with intervals of low to no recovery in between. Continuous coring was made impossible by extremely harsh weather conditions and fast-moving sea ice at the time of drilling. The core units are consecutively named core NBP0602A-3C-1 from the top of the section to core NBP0602A-3C-7 at the base. Figure 2 presents a summary of the zone of recovery and the relationship between each unit with respect to depth below seafloor [*Bohaty et al.*, this volume]. By convention, recovered core has been placed at the top of each cored interval. This plate also presents a summary of the onboard diatom zonal assignment and diatom age determination [*Bohaty et al.*, this volume]. Based on preliminary diatom biostratigraphy, units in core NBP0602A-3C-7 to most of core NBP0602A-3C-1 were assigned a maximum age of 37 Ma and a minimum age of 32 Ma, with, disconformably overlying them, a thin (less than 1 m) upper Pleistocene- to Holocene-recovered section marking the top of core NBP0602A-3C-1. Sample NBP0602A-3C-1, containing 100% reworked polynomorphs, is not discussed. Based on the diatom age, it was confirmed that the section sampled was therefore uppermost Eocene to lowermost Oligocene.

Based on their seismic stratigraphic analysis, *Anderson et al.* [2006] noted that, given the great thickness of the total sediment package present at this site, the ~10 m thick section sampled in Hole NBP0602A-3C must only represent a very short interval of time within the assigned age range. Furthermore, the sediments from this hole mostly consist of muddy to very fine sand that varies in color from greenish black in the upper portion of the hole (0–7.5 mbsf) to very dark greenish gray in the lower portion (7.5–20.0 mbsf) of the core [*Anderson et al.*, 2006]. So, it is assumed that the section recovered is essentially continuous, but a minor hiatus might be present around 7.5 mbsf.

Twenty samples were collected for palynological analysis. Location of palynological samples relative to the various core units recovered is indicated in Figure 2. As marked in this plate, two to three samples were taken at regular intervals in each of the cored units recovered from NBP0602A-3C-1 to NBP0602A-3C-7. The detailed location of these samples can be found in Table 1.

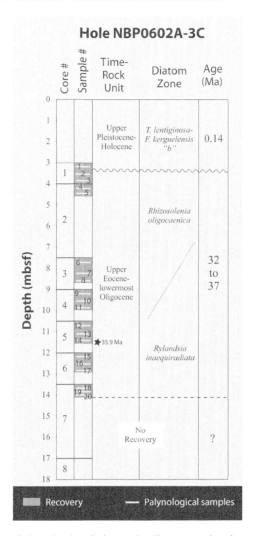

Figure 2. Palynological sample location in relation to the diatom zonal assignments and age interpretation for Borehole NBP0602A-3C. Note that sample 1 is of upper Pleistocene-Holocene age. Modified after Figures 4–6 of the Shipboard Scientific Party (SHALDRIL II 2006 NBP0602A cruise report, 2006, available at http://www.arf.fsu.edu/projects/documents/Shaldril_2_ Report.pdf, hereinafter referred to as Shipboard Scientific Party, 2006).

These samples were chemically processed following a palynological technique based on a mix of traditional methods such as those summarized by *Brown* [2008]. For each sample, about 10 g of dried sediments were weighed (the precise weight for each sample is listed in Table 1). This weight is important as it is used to calculate palynomorph concentration per gram of dried sediment to allow evaluation of changes in concentration throughout the core. The sediment was then spiked with a known quantity of *Lycopodium* spores to allow computation of the absolute abundance of palynomorphs in the sample. Acid soluble minerals were digested in HCl and HF acid to remove carbonates and silicates. The palynomorphs were then concentrated by filtration on a 10 μm mesh sieve. Samples were sufficiently rich to allow the tabulation of 300 palynomorphs per sample. A database of all palynomorphs recovered was prepared, and key species were photographically documented.

Table 1. Details of Sampled Core Sections of NBP0602A-3C

Sample	Depth of Sample per Core Section (cm)	Weight (g)	Dinocyst Concentration	Pollen Concentration (specimen per gram of dried sediments)
		Core Section NBP0602A-3C-1		
1	14–16	13.4	23300	700
2	54–56	15	1800	2,100
3	88–90	15	2400	1,100
		Core Section NBP0602A-3C-2		
4	10–12	11.6	4800	2,200
5	29–31	10	16300	3,800
		Core Section NBP0602A-3C-3		
6	11–13	10	9100	10,800
7	59–61	10.1	9800	8,100
8	95–97	10.1	3000	3,300
		Core Section NBP0602A-3C-4		
9	10–12	10	8400	6,100
10	49–51	10	6600	30,800
11	85–87	10	4800	3,500
		Core Section NBP0602A-3C-5		
12	10–12	10	11800	13,900
13	54–56	10	18000	19,000
14	81–83	6.5	18700	33,300
		Core Section NBP0602A-3C-6		
15	10–12	10	16700	14,200
16	49–51	8	92000	100,000
17	83–85	8	55000	44,700
		Core Section NBP0602A-3C-7		
18	6–8	8	26600	33,500
19	29–31	8	30600	21,700
20	57–59	8	103300	75,300

Repository and curation of all palynological slides will be handled by the Louisiana State University Center for Excellence in Palynology (CENEX).

3. PALYNOLOGICAL RESULTS

3.1. General Evaluation

Well-preserved penecontemporaneous terrestrial and marine palynomorphs, as well as recycled palynomorphs, were recovered. The concentration of dinoflagellate cysts per gram of dried sediments (d/g) and pollen and spores were calculated and graphically represented (Figure 3). The

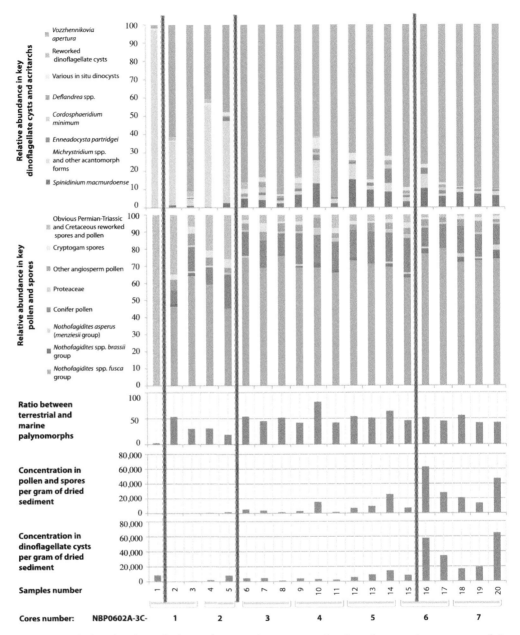

Figure 3. Relative abundance in key palynomorphs, concentration in palynomorphs per gram of dry sediment and ratio between terrestrial and marine palynomorphs, related to sample and core numbers.

concentrations of dinoflagellate cysts per gram of dried sediments (d/g) range from a high maximum of 64,800 d/g in the bottom of the core to a minimum 600 d/g in the top of the core. The concentrations of pollen and spores per gram of dried sediments (p/g) range from quite good 47,200 p/g in the bottom of the core to a low 700 p/g in the top of the core, with a maximum of 62,700 p/g in sample 16 (NBP0602A-3C-6-49) and a minimum of 400 p/g in sample 3 (NBP0602A-3C-1-88).

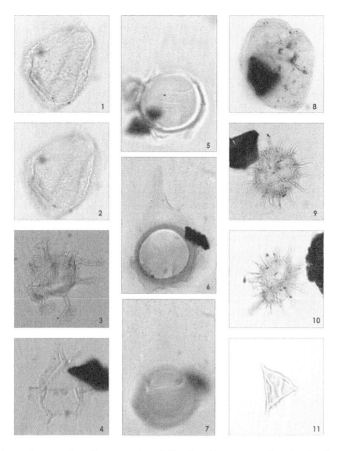

Figure 4. Light photomicrographs of key species of dinoflagellate cyst and acritarch. For each sample, the sample number (e.g., NBP0602A-3C-1) is followed by the sample depth (e.g., 0.54) and then by the England Finder coordinates of each specimen. Numbers represent the following: 1 and 2, *Vozzhennikovia apertura*, NBP0602A-3C-1.54, V41/3; 3, *Enneadocysta partridgei*, NBP0602A-3C-5.10, T43/1; 4, *Spinidinium macmurdoensis*, NBP0602A-3C-1.54, U21/4; 5, *Deflandrea antarctica*, NBP0602A-3C-4.49, P42/3; 6 and 7, *Deflandrea cygniformis*, NBP0602A-3C-4.85, Q29/4; 8, *Manumiella seymourensis*, NBP0602A-3C-1.54, P41/4; 9, *Micrhystridium* sp., NBP0602A-3C-1.14, V33/4; 10, *Impletosphaeridium*? sp. (note archeopyle), NBP0602A-3C-1.14, U27/4; and 11, *Veryhachium reductum*, NBP0602A-3C-3.59, N44/3.

Interestingly, after relatively constant high concentrations of both marine and terrestrial paly-nomorphs from samples 20 through 16, a sharp decrease in concentration is observed, with concentrations remaining low from samples 15 to 2 (Figure 3). This significant sudden drop in concentration of ~85% for marine palynomorphs and of ~90 % in terrestrial palynomorphs could reflect a sharp decline in plant and phytoplankton productivity resulting from a major shift toward colder climatic conditions, felt both on land and in the marine realm. But this signal could also be induced by a higher sedimentation rate that would give the appearance of lower plant and phytoplankton productivity, while simply associated with increased terrigenous input.

3.2. Dinoflagellate Cysts and Acritarchs

The relative abundances of key dinoflagellate and acritarch species are presented in Figure 3. Key dinoflagellate and acritarch species are illustrated in Figure 4. The marine palynomorph assemblage in most samples is dominated by *Vozzhennikovia apertura*. That species is relatively frequent throughout the La Meseta Formation on Seymour Island [*Wrenn and Hart*, 1988], especially at Cape Wiman [*Cocozza and Clarke*, 1992]. It has also been reported as a common component of the assemblage in the Eocene Southern Ocean [*Sluijs et al.*, 2003; *Brinkhuis et al.*, 2003]. However, in all instances, none of these authors reported such dominance in *V. apertura* in the assemblage. Other species recorded in NBP0602A-3C borehole are *Enneadocysta partridgei*, *Deflandrea antarctica* along with specimens of *Deflandrea cygniformis* and *Deflandrea phosphoritica*, *Spinidinium macmurdoensis*, and *Cordosphaeridium minimum*. *Vozzhennikovia rotunda* was found in a few samples but only as rare occurrences. All of these are typical high southern latitude taxa known as members of the "Transantarctic Flora" [*Wrenn and Beckmann*, 1982], but again, the relative abundances in Borehole NBP0602A-3C are quite different as *V. apertura* clearly dominates the assemblage here.

Particularly notable is the occurrence in the uppermost Eocene sampled at 3C (samples 6 and above) of common reworked dinoflagellates of Cretaceous age, such as *Isabelidinium cretaceum*, and various species of the genus *Manumiella*, including *Manumiella seymourensis*. These peridinioid dinoflagellates are the most frequently occurring reworked Cretaceous forms, "after" the abundant acanthomorphs discussed below. Rare specimens of other taxa such as *Odontochitina operculata*, *Palaeocystodinium* sp., and various chorate forms were also noted. Only occasional specimens occur in lower samples.

In samples 5 and above, these reworked Cretaceous dinoflagellate cysts are associated with a major increase in diverse acanthomorph acritarchs such as *Micrhystridium* spp. and occasional *Impletosphaeridium* spp. Most of these forms are darker in color than presumed in-place clear/translucent forms, indicating a thermally more mature status. The thermally mature *Micrhystridium* almost all have a golden hue, compared to the translucent, sometimes almost invisible small forms (more common in lower cores) with definite capitate process tips that are included in the genus *Impletosphaeridium*. The presence of an archeopyle allows for inclusion of some acanthomorph-like forms in the dinocyst *Impletosphaeridium* genus, but an archeopyle is not always easily identifiable. Furthermore, many variations have been observed in the process shape (thinner versus stouter process base, flexible versus straighter process) among the various small acanthomorph forms that we grouped as *Micrhystridium* spp. The higher degree of thermal maturity and the association of these acanthomorph forms with the definite Cretaceous (Campanian-Maastrichtian) dinoflagellate cysts such as *Isabelidinium cretaceum* and *Manumiella seymourensis* in our view confirms that the extremely abundant *Micrhystridium* spp. found in the upper part of NBP0602A-3C are reworked. *Askin* [1988, 1999] noted that "swarms" or abundant *Micrhystridium* spp. were associated with *I. cretaceum* and the lower range of *M. seymourensis* on Seymour Island. This confirms erosion and redeposition of upper Campanian-lower Maastrichtian rocks in our studied sediments. These Cretaceous forms are found throughout the cored interval, but become abundant in samples 5 and above, indicating that downcutting and erosion is significantly increased at that time.

3.3. Pollen and Spores

Based on the concentration observed and composition (with moderate diversity) of spore and pollen assemblages, the terrestrial palynomorphs indicate that, at the time of deposition for most of the cores, southern beech-dominated vegetation was abundant on the adjacent landmass. It is, however, not always easy to differentiate between penecontemporaneous and recycled origins for much of the assemblage. Some specimens are obviously recycled, as evidenced by their somewhat darker color (higher thermal maturation) and morphologies (e.g., taeniate bisaccate pollen

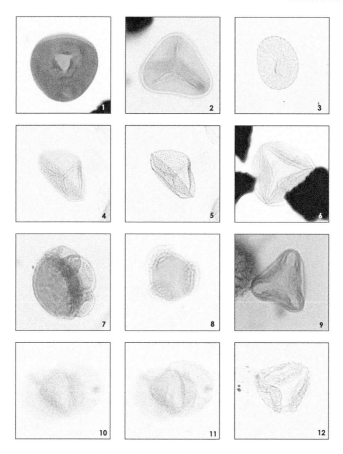

Figure 5. Light photomicrographs of selected species of spore and gymnosperm pollen. For each sample, the sample number (e.g., NBP0602A-3C-1) is followed by the sample depth (e.g., 0.54) and then by the England Finder coordinates of each specimen. Numbers represent the following: 1, unknown spore, NBP0602A-3C-1.54, P33/1; 2, *Cyathidites minor*, NBP0602A-3C-4.85, T43/2; 3, *Coptospora* sp., NBP0602A-3C-6.49, T31/1; 4 and 5, *Osmundacidites wellmanii/Baculatisporites comaumensis* complex, NBP0602A-3C-4.10, R28/2; 6, *Podosporites/Trichotomosulcites* complex, NBP0602A-3C-1.54, U45/3; 7, *Microcachryidites antarcticus*, NBP0602A-3C-1.54, U39/4; 8, *Microcachryidites* cf. *antarcticus*, NBP0602A-3C-7.57, W35/1; 9, *Podosporites/Trichotomosulcites* complex, NBP0602A-3C-4.85, R36/1; 10 and 11, *Podocarpidites* sp., NBP0602A-3C-4.85, Q23/3; and 12, *Podosporites/Trichotomosulcites* complex, NBP0602A-3C-4.10, S33/1.

characteristic of the Permian). For this reason, the designation in Figure 3 for recycled pollen and spores is labeled "Obvious." Numbers within the recycled group are likely underestimated, however, and the other "in-place" categories likely include recycled specimens that are of similar thermal maturation to the penecontemporaneous flora and have an age range elsewhere that does not with certainty distinguish them as recycled. Some ambiguous occurrences are noted below. Similar problems were encountered in cores recovered from the Ross Sea (e.g., Cape Roberts Project [*Raine*, 1998; *Askin and Raine*, 2000; *Raine and Askin*, 2001]).

Selected spores and gymnosperm pollen are illustrated in Figure 5, and angiosperm pollen are shown in Figure 6. The terrestrial assemblage is conspicuously dominated by pollen of the *Nothofagidites fusca* group, including at least 10 species (some of which may be recycled, as

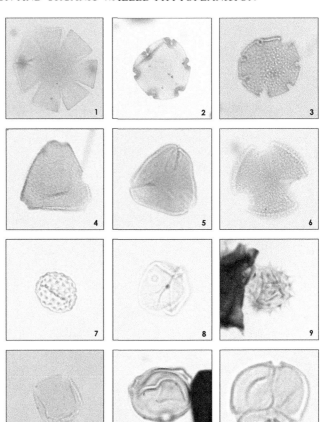

Figure 6. Light photomicrographs of selected species of angiosperms. For each sample, the sample number (e.g., NBP0602A-3C-1) is followed by the sample depth (e.g., 0.54) and then by the England Finder coordinates of each specimen. Numbers represent the following: 1, *Nothofagidites asperus*, NBP0602A-3C-4.85, Q36/2; 2, *Nothofagidites* sp. cf. *N. flemingii/N. saraensis* complex, NBP0602A-3C-1.54, V49/4; 3, *Nothofagidites lachlaniae*, NBP0602A-3C-1.54, V30/1; 4, *Proteacidites pseudomoides*, NBP0602A-3C-1.54, R41/1; 5, *Tricolpites* ?*gillii*, NBP0602A-3C-1.54, U28/3; 6, tricolpate reticulate pollen, NBP0602A-3C-4.85, T25/4; 7, Chenopodiaceae, NBP0602A-3C-2.29, T23/3; 8, Poaceae, NBP0602A-3C-3.59, T41/4; 9, Asteraceae, NBP0602A-3C-1.88, Q43/2; 10, Stylidiaceae, NBP0602A-3C-2.29, R15/3; 11, *Myricipites harrisii*, NBP0602A-3C-1.14, Q46/3; and 12, syncolporate pollen, NBP0602A-3C-3.11, U34/1.

noted above). They include *Nothofagidites lachlaniae*, species similar to (or assignable to) *Nothofagidites waipawaensis* and *Nothofagidites brachyspinulosus* types, the complex of *Nothofagidites flemingii*-cf. *flemingii* (a smaller form, most common of this complex, as in the younger Ross Sea assemblages)- *Nothofagidites saraensis*, and other types such as the moderately large form similar to *N. flemingii* but lacking colpal thickenings and provisionally listed as *N.* sp. A in Ross Sea assemblages. These "*fusca* group" pollen types have similarities to pollen from both the extant *Nothofagus* (southern beech) subgenera *Fuscospora* and *Nothofagus*. There are also scattered occurrences, slightly more common in the lower samples, of *Nothofagidites* spp. *brassii* group (extant beech subgenus *Brassospora*), and *Nothofagidites asperus* (or similar forms, *menziesii* group, extant beech subgenus *Lophozonia*). The overwhelming abundance of *N. fusca* group pollen may be in part related to these pollen being windblown and produced in huge quantities (also true for podocarp conifer pollen, which are also quite common in the NBP0602A-3C cores), but it is believed that the adjacent vegetation was likely *Nothofagus*-dominated with a high relative diversity of southern beech species. These, and the conifers described below, might have included forest trees to stunted shrubs in more exposed locations and near tree lines.

The second largest grouping within the terrestrial component is the conifer pollen. Almost all of these are podocarp conifer pollen, but there are also a few *Araucariacities australis*, and rare possible Cupressaceae/Taxodiaceae (though similar to some leiosphaerid type algae). The podocarp conifer pollen are quite varied and can mostly be grouped in the bisaccate genera *Podocarpidites* and *Phyllocladidites-Microalatidites* types, with less common but variable pollen of the trisaccate *Podosporites-Trichotomosulcites* type. A few specimens of *Phyllocladidites mawsonii* have been observed, but they may well be recycled (these are common in Cretaceous sediments from the James Ross Basin, as are the trisaccate types), along with *Dacrydiumites* spp. and *Microcachrydites antarcticus*. We note that there are also specimens similar to *P. mawsonii* but without the characteristic boss at the base of each sac and specimens similar to *M. antarcticus* but smaller and with relatively smaller sacs. Even discounting some recycling, there appears to have been a good diversity of podocarp conifers on the adjacent land mass.

Rare Proteaceae pollen are mostly small simple types assignable or similar to *Proteaceidites parvus* and *Proteaceidites subscabratus*, plus rare occurrences of other species such as *Proteaceidites pseudomoides*, *Beaupreaidites* spp. and *Granodiporites* sp. Other angiosperm pollen include rare *Liliacidites*, Restionaceae (monoporate and scrobiculate), and Poaceae (more strongly annulate and smooth, not modern contaminants), various small simple mostly nondescript tricolpate, tricolporate, and triporate pollen, and some larger more distinctive triporate pollen also known from Seymour Island Eocene such as *Myricipites harrisii*, plus rare Caryophyllaceae/Chenopodiaceae types, Ericales, Stylidiaceae, Droseraceae (*Fischeripollis* sp.), and Asteraceae (the latter appear not to be contaminants). Cryptogam spores are rare to infrequent and not very diverse.

The pollen and spores recovered are indicative of a vegetative cover on nearby land similar to that recorded from the La Meseta Formation on Seymour Island, although the SHALDRILL II uppermost Eocene record shows a lower-diversity plant association than that found in the lower and middle La Meseta Formation [*Askin*, 1997], with *Nothofagidites* spp. *fusca* group clearly dominating the flora and a secondary varied podocarp conifer component. A shift to *fusca*-dominated *Nothofagidites* floras, with concomitant loss of varied *N. brassii* group species occurred in the upper La Meseta Formation [*Chen*, 2000; R. Askin, unpublished data, 2010]. Unfortunately, a meaningful comparison cannot be made with the uppermost La Meseta assemblage as those sediments were very sandy, and it was believed that small *Nothofagidites* pollen might have been winnowed out with the mud fraction during transport and deposition [*Askin*, 1997].

There are some similarities to younger Ross Sea Oligocene and Miocene assemblages, including many of the same *Nothofagidites* and podocarp conifer pollen species, and most notably some components of the "tundra assemblage" of ANDRILL-2A [*Warny et al.*, 2009] and the Cape Roberts Project [*Askin and Raine*, 2000; *Raine and Askin*, 2001]. These include *Coptospora* spp., Caryophyllaceae/Chenopodiaceae types, Poaceae, Asteraceae, and Stylidiaceae. Although these are only rare occurrences in 3C, it is of interest to note the appearance of these at the end of the Eocene in this area. They are present in the middle and upper cores above sample 16.

Similar aged late Eocene assemblages, though substantially more diverse, especially among Proteaceae, have recently been described from Prydz Bay [*Truswell and Macphail*, 2009]. Although recycling may account for some of the diversity, these authors instead suggest certain taxa may have survived longer in the cooler East Antarctic environments than in lower-latitude locales such as Australia where their last known occurrences do not extend into the Eocene. For these late Eocene Prydz Bay floras, a mosaic vegetation of dwarfed (krumholtz) trees, sclero-morphic shrubs, and wetland herbs, analogous to the present northern taiga, was suggested [*Truswell and Macphail*, 2009]. The Eocene Proteaceae on the Antarctic Peninsula were apparently much less diverse and suggest the scleromorphic component of the flora might be lacking. This is consistent with a much higher rainfall/higher humidity climate in the Eocene Antarctic Peninsula, (similar to present-day southern Chile and southwestern New Zealand), as might be expected in the more maritime peninsula conditions, compared to coastal East Antarctica. The taiga-like vegetation of dwarfed trees envisaged for the similar aged Prydz Bay floras is a good analogy for the interval represented by the middle and upper NBP0602A-3C cores, with tundra developing on the higher and more exposed parts of the landscape.

In addition to in situ pollen and spores, recycled Cretaceous forms (e.g., *Cicatricosisporites australiensis*), many with a darker hue and battered and corroded so that many are unidentifiable to species level, are found scattered throughout the cores. They become substantially more common in the uppermost section, starting at sample 5. A similar trend is observed with the marine palynomorphs.

Permian-Triassic recycled specimens are found sporadically throughout the core, though are somewhat more common in the lower samples. Permian-Triassic pollen and spores occur throughout the Seymour Island succession as well but are noticeably more common in the upper La Meseta Formation [*Askin and Elliot*, 1982], a further similarity to the lower NBP0602A-3C cores. Like the La Meseta recycled specimens, those found throughout 3C mostly exhibit relatively low thermal maturation with their yellow-orange to orange-brown coloration, compared to brown-black Permian palynomorphs found in place in the Transantarctic Mountains. Relatively unaltered emergent forearc sediments adjacent to the Antarctic Peninsula were suggested as a possible source for the recycled palynomorphs by *Askin and Elliot* [1982].

4. DISCUSSION

4.1. Biostratigraphy and Global Implications of the Age Model

Stratigraphic inferences for Hole NBP0602A-3C are mainly based on a combination of seismic profiles tied to diatom and nannofossil biostratigraphy. These biostratigraphic data are presented in the SHALDRIL II shipboard report (Shipboard Scientific Party, 2006) and are summarized by *Bohaty et al.* [this volume]. Although dinoflagellate cysts, as yet, are only of limited use in Antarctic Neogene biostratigraphic studies, some promising work has been published recently for some Oligocene-Neogene sections [*Hannah et al.*, 2000, 2001a, 2001b] showing that dinoflagellate cysts are good stratigraphic markers for Cretaceous to Eocene Antarctic sections [e.g., *Wrenn and Hart*, 1988; *Hall*, 1977]. The dinoflagellate cyst biostratigraphic model in this study allows us to narrow the age of the drilled section. Based on the presence of the diatoms *Rhizosolenia oligocaenica* and *Rylandsia inaequiradiata* [*Bohaty et al.*, this volume], the age of the section was estimated to range anywhere between 32 and 37 Ma. Based on known distributions of key dinoflagellate cysts present in the NBP0602A-3C borehole, mainly the last occurrence datum of *Spinidinium macmurdoense* and *Vozzhennikovia rotunda*, this window can be narrowed to 33.7 to 37 Ma [*Bohaty et al.*, this volume]. This age is based on dinoflagellate cyst ranges summarized by *Williams et al.* [2003], based on their review of literature and new data from Ocean Drilling Program Leg 189 in the Tasmanian seaways. This biostratigraphic interpretation is further supported by the 35.9 Ma strontium date obtained at the base of core NBP0602A-3C-5 [*Bohaty et al.*, this volume]. These combined stratigraphic data confirm that Borehole NBP0602A-3C sampled the Priabonian.

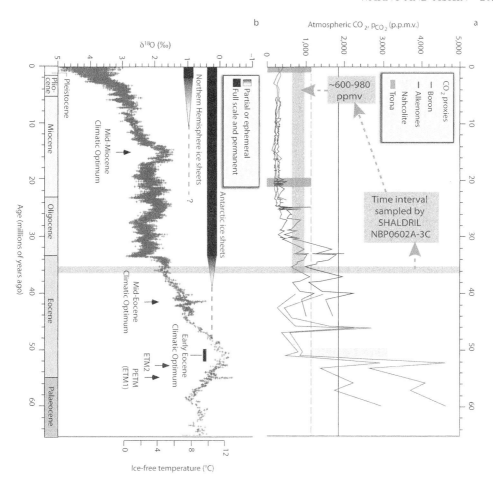

Figure 7. Stratigraphic position of sediment recovered from the NBP0602A-3C borehole in relation to the atmospheric CO_2 levels and global climate. The horizontal shaded zone represents the maximum interval of time sampled by NBP0602A-3C (this is the maximum range possible based on the biostratigraphy available for the well). The central vertical shaded zone projects the width of the atmospheric CO_2 curve at the time sampled by NBP0602A-3C on the atmospheric CO_2 axis to give a minimum and maximum range of atmospheric CO_2 value at the time of deposition. Modified from *Zachos et al.* [2008]. Reprinted by permission from Macmillan Publishers Ltd: *Nature*, copyright 2008 (http://www.nature.com/nature/index. html).

Comparison with the atmospheric CO_2 levels and oxygen isotope proxy curve presented in the *Zachos et al.* [2008] summary (Figure 7) indicates that NBP0602A-3C recovered a key stratigraphic interval in the Priabonian. Although the sedimentary layers sampled cannot be more precisely dated than being deposited sometime between 37 and 33.7 Ma, it can be inferred that the palynological record from this interval captured the response of plants and organic-walled phytoplankton during the first marked increase in $\delta^{18}O$ in the latest Eocene, followed by a large reduction in atmospheric $CO_2.$ This event directly precedes the major Oi-1 glaciation that marks the Eocene/Oligocene boundary. During the interval sampled by SHALDRIL II, the atmospheric CO_2 could have been as low as 600 ppmv but up to 980 ppmv depending on the various CO_2

proxies presented by *Zachos et al.* [2008]. To put these values into the context of modern climate debates, the concentration of carbon dioxide in the atmosphere (based on data available on the NASA Web site) at the start of the industrial revolution was about 280 ppmv; it then rose to 300 ppmv by the late 1940s, reaching today's concentration of 360 ppmv. At these current rates, the atmospheric CO_2 is expected to reach 560 ppmv, within the next 100 years. If the minimum values of atmospheric CO_2 presented in the *Zachos et al.* [2008] summary represent the environmental atmospheric concentration at time of NBP0602A-3C deposition, this means that the system and vegetative cover represented by the NBP0602A-3C cored interval might reflect what can be expected for the next century.

4.2. Discussion on Ecology and Assemblage Composition

Overall, the low-diversity/high-dominance dinoflagellate cyst assemblage is clearly a sign of deteriorating conditions compared to those observed in most of the Eocene section sampled on Seymour Island. In the NBP0602A-3C section, the majority of species present throughout the interval are peridinioid species (*Vozzhennikovia* spp., *Deflandrea* spp., *Spinidinium* spp.). If these species are indeed heterotrophic, as is the case for many modern peridinioids, then their dominance suggests that surface waters were eutrophic at the time these algae lived. This could indicate an environment such as an upwelling zone or a deltaic complex. Most of these dinoflagellate cysts are common in the Transantarctic Flora as described by *Wrenn and Hart* [1988], but the high relative abundance in *V. apertura*, a thicker, possibly more robust dinoflagellate cyst, might also illustrate the fact that the environment has a higher energy, possibly as a result of the general shallowing of the section, though other lines of evidence are lacking for this hypothesis. It is more likely that the low-diversity/high-dominance assemblage signifies significantly colder surface waters compared to the environments known from the Seymour Island Eocene sections.

Throughout the interval, both marine and terrestrial species recovered indicate cold climatic conditions. For the lower cores, the vegetation might have grown in climatic conditions similar to those found in the colder forested parts below tree line of southern Chile and southwestern New Zealand today and somewhat harsher than those suggested for the middle Eocene of Seymour Island. *Francis et al.* [2008, and references therein], based on analysis of leaves, suggested a markedly seasonal climate for the middle Eocene of this region with a mean annual temperature of 10.8°C ± 1.1°C, a warm month mean of 24°C ± 2.7°C, a cold month mean of −1.17°C ± 2.7°C, and 1534 mm annual rainfall.

Environmental conditions both on land and in the marine realm seem to have taken a turn for the worse between the deposition of samples 16 to 15. At that level, the high concentration in palynomorphs observed in the lower cores (samples 16 through 20) dropped by 85% for marine palynomorphs and by 90% for terrestrial palynomorphs and remained low in the rest of the cored interval. The change in concentration is also paralleled by relative abundance increases in pollen of conifers and the dinoflagellate cyst *V. apertura*. But this drop may be a reflection, at least in part, of sudden decrease in marine and terrestrial palynomorph productivity, although changes such as an increased sedimentation rate cannot be ruled out. However, no obvious changes in the lithology were reported between samples 16 and 15 where the concentration drop is observed, so a possible change in sedimentation conditions and possibly provenance are not well supported. Furthermore, one would expect that the important increase in terrigenous input indicative of higher sedimentation rate might be associated with input of reworked palynomorphs into the assemblage, but such an increase is not observed in sample 15. Thus, the lower-productivity hypothesis is favored over the increased sedimentation rate. *Diester-Haass and Zachos* [2003] evaluated the change in paleoproductivity during the Eocene-Oligocene transition. Their record indicates a prominent increase in productivity in the Southern Ocean. They postulated that this widespread increase may have led to enhanced carbon burial that contributed to CO_2 extraction, hence acting as positive feedback to the already cooling climate. Increases in productivity are not seen in any of our younger Eocene samples. This suggests that the increased productivity discussed by Diester-Haass and Zachos was not the result of vegetation or organic-wall plankton proliferation, but most likely, a consequence of blooms in siliceous-walled microfossil microplankton such as radiolarians or diatoms, which we know thrive today in polar and subpolar conditions in response to abundance in food and silica content in polar and subpolar ocean waters.

Considering this drastic decrease in concentration in sample 15 and above, it does appear that conditions changed simultaneously on land and in the ocean, as there is no discernible lag in major assemblage composition when marine and terrestrial assemblages are compared. Further detailed species by species comparisons in both marine and terrestrial assemblages will be needed to confirm this initial interpretation. The concentrations in palynomorphs keep decreasing substantially toward the middle of the section, possibly linked to further deteriorating conditions, and decrease even more at the top of the section. Unlike elsewhere in the section, this upper decrease is coincident with a significant increase in reworked palynomorphs. This major change may record a more significant sea level drop, induced by substantial cooling and development of glaciers. Based on the age of the majority of the reworked palynomorphs, the reworked marine and terrestrial forms were likely incorporated in the sediment during downcutting and redeposition of nearby Campanian-Maastrichtian sections at the end of the Eocene.

Oxygen isotopic values for the time interval recovered by SHALDRIL indicate an increasing trend. Based on the oxygen isotope record alone, it cannot be distinguished if these are characteristics of increased ice volume, decreasing temperature, or both. On the basis of the SHALDRIL NBP0602A-3C palynological record, a two-step cooling is suggested. We propose that during the first step, cooling caused the drop in productivity and thus concentration in the preserved palynomorph assemblage between samples 16 and 15. The second cooling step occurred between samples 6 and 5 and is marked by greatly increased reworking activity, hence indicating a drop in sea level most likely related to ice sheet expansion. This downcutting event is unfortunately not dated with great precision, although evidence shows it occurred between 33.7 and 35.9 Ma. During this time interval, around 35.5 Ma, a marked decrease in temperature is seen in the oxygen isotope curve and in the TEX_{86} results of DSDP Site 511 off Argentina [*Liu et al.*, 2009]. For that time interval, both the oxygen isotopes and the TEX_{86} method indicate a drop in temperature of 6°C to 7°C. Although these methods cannot indicate the presence of an ice sheet, this cooling event might be correlated to the major sea level drop we report between samples 6 and 5. Interestingly, the replacement of prodeltaic endemic Antarctic dinoflagellate cyst assemblages by open marine cosmopolitan communities at three Ocean Drilling Program Leg 189 sites (1170, 1171, and 1172) indicates that the deepening of the Tasmanian Gateway was quasi-synchronous throughout the Tasmanian region and that it started at ~35.5 Ma [*Sluijs et al.*, 2003].

The second cooling event that we report between the deposition of cores NBP0602A-3C-3 (sample 6) and NBP0602A-3C-2 (sample 5) appears to be associated with ice sheet expansion and is dated between 35.9 and 33.7 Ma. Although the palynological changes indicative of sea level drop observed between samples 6 and 5 are abrupt, it is possible that these changes actually happened gradually. Stratigraphic data do not have the resolution needed to identify whether the sequence is continuous or if a minor hiatus marks this level. Actually, a noticeable change in sediment coloration, from dark greenish gray for samples 6 and below to greenish black for samples 5 and above [*Anderson et al.*, 2006], might indicate that such a hiatus indeed exists between samples 6 and 5.

From the NBP0602A-3C record, we see that both dinoflagellate assemblages and the vegetation indicate cooler climatic conditions sometime between 35.9 and 33.7 Ma. Without more precise stratigraphic resolution, it is not possible to determine whether the changes we see in the Weddell Sea occurred concurrently with the 35.5 Ma deepening of the Tasmanian Gateway, but we argue that the lowering of atmospheric CO_2 concentration must have played the predominant role for the cooling trend seen prior to 35.5 Ma.

5. CONCLUSIONS

Significant global decrease in diversity of organic-walled phytoplankton and in vegetation had already occurred by the end of the Eocene, at least by 36 Ma, hence at least 2.3 Myr before the sharp marine oxygen isotope ratio shift that reflects marked global cooling at the Eocene-Oligocene boundary and before the 35.5 Ma deepening of the Tasmanian Gateway. This is clearly indicated by the low diversity/high abundance in a few species of dinoflagellate cysts such as *V. apertura*, *S. macmurdoense*, and *D. antarctica* and in the dominance, among the vegetation, of pollen of *Nothofagidites* spp. *fusca* group with secondary podocarp conifers.

Following this cooling phase, a second, possibly more intense cooling took place in the upper part of the section. In these upper samples, in addition to extremely low concentration of in situ palynomorphs, a significant increase in reworking activity is clearly obvious. The abundance in the acritarch *Micrhystridium* spp. along with significant increase in reworked Cretaceous dinoflagellate cysts, pollen, and spores at that level is considered to represent a period of intense erosion of nearby upper Campanian–lower Maastrichtian sections. This downcutting was most likely induced by a marked sea level drop, probably associated with growth of ice sheets in and off Antarctica. This may be correlated to what is interpreted in the seismic profile as a major onlap surface, which *Smith and Anderson* [this volume] argued represents a significant low-stand. It is possible (D. H. Elliot, personal communication, 2010) that this surface corresponds to the erosion surface on top of the meseta on Seymour Island, between the La Meseta Formation and the overlying ?Eocene/Oligocene glacial beds described by *Ivany et al.* [2006]. A possible relationship is also suggested to some of the valley glaciers that are reported from King George Island around that time [*Francis et al.*, 2009; J. B. Anderson, personal communication, 2010].

Based on biostratigraphy and the 35.9 Ma strontium date obtained at the base of core NBP0602A-3C-5, it is postulated that the drop in concentration in the lower part of the core and major lowering of sea level in the upper part of the core are most likely related to the brief spike in $\delta^{18}O$-enriched value observed from 36 to 35.5 Ma in the *Zachos et al.* [2001, 2008] composite record. If this date is correct, the section recovered by SHALDRIL gives us a glimpse into what the world might look like once carbon dioxide concentrations in the atmosphere reach a level two to three times our current value.

Acknowledgments. Thanks are extended to the National Science Foundation for funding this research (grant from NSF Office of Polar Program 0636747) and to John Anderson of Rice University and Julia Wellner of the University of Houston for providing the samples necessary for the completion of this study. Thanks are also extended to the Antarctic Research Facility staff for giving us access to the cores and assisting us with sampling needs.

REFERENCES

Anderson, J. B. (1999), *Antarctic Marine Geology*, 297 pp., Cambridge Univ. Press, Cambridge, U. K.

Anderson, J. B., J. S. Wellner, S. Bohaty, P. L. Manley, and S. W. Wise Jr. (2006), Antarctic Shallow Drilling Project provides key core samples, *Eos Trans. AGU*, *87*(39), 402.

Anderson, J. B., et al. (2011), Progressive Cenozoic cooling and the demise of Antarctica's last refugium, *Proc. Natl. Acad. Sci. U. S. A.*, *108*(28), 11,356–11,360, doi:10.1073/pnas.1014885108.

Askin, R. A. (1988), Campanian to Paleocene palynological succession of Seymour and adjacent islands, northeastern Antarctic Peninsula, *Mem. Geol. Soc. Am.*, *169*, 131–154.

Askin, R. A. (1997), Eocene - ?earliest Oligocene terrestrial palynology of Seymour Island, Antarctica, in *The Antarctic Region: Geological Evolution and Processes*, edited by C. A. Ricci, pp. 993–996, Terra Antart. Publ., Siena, Italy.

Askin, R. A. (1999), *Manumiella seymourensis* new species, a stratigraphically significant dinoflagellate cyst from the Maastrichtian of Seymour Island, Antarctica, *J. Paleontol.*, *73*(3), 373–379.

Askin, R. A., and D. H. Elliot (1982), Geologic implications of recycled Permian and Triassic palynomorphs in Tertiary rocks of Seymour Island, Antarctic Peninsula, *Geology*, *10*(4), 547–551.

Askin, R. A., and J. I. Raine (2000), Oligocene and early Miocene terrestrial palynology of the Cape Roberts drillhole CRP-2/2A, Victoria Land Basin, Antarctica, *Terra Antart.*, *7*(4), 493–501.

Bohaty, S. M., D. K. Kulhanek, S. W. Wise Jr., K. Jemison, S. Warny, and C. Sjunneskog (2011), Age assessment of Eocene–Pliocene drill cores recovered during the SHALDRIL II expedition, Antarctic Peninsula, in *Tectonic, Climatic, and Cryospheric Evolution of the Antarctic Peninsula*, doi:10.1029/2010SP001049, this volume.

Brinkhuis, H., S. Sengers, A. Sluijs, J. Warnaar, and G. L. Williams (2003), Latest Cretaceous-earliest Oligocene and Quaternary dinoflagellate cysts, ODP Site 1172, east Tasman Plateau [online], *Proc. Ocean Drill. Program Sci. Results*, *189*, 48 pp. [Available at http://www-odp.tamu.edu/publications/189_SR/106/106.htm.]

Brown, C. (2008), *Palynological Techniques*, edited by J. Riding and S. Warny, Am. Assoc. of Stratigr. Palynol. Found., Dallas, Tex.

Chen, B. (2000), Evolutionary, biogeographic and paleoclimatic implications of Eocene *Nothofagidites* pollen from Seymour Island, Antarctica, M.S. thesis, Ohio State Univ., Columbus.

Cocozza, D., and C. M. Clarke (1992), Eocene microplankton from La Meseta Formation, northern Seymour Island, *Antarct. Sci.*, *4*, 355–362.

DeConto, R. M., and D. Pollard (2003), Rapid Cenozoic glaciation of Antarctica induced by declining atmospheric CO_2, *Nature*, *421*, 245–249.

Diester-Haass, L., and J. Zachos (2003), The Eocene-Oligocene transition in the equatorial Atlantic (ODP Site 925): Paleoproductivity increase and positive $\delta^{13}C$ excursion, in *From Greenhouse to Icehouse: The Marine Eocene-Oligocene Transition*, edited by D. R. Prothero, L. C. Ivany, and E. R. Nesbitt, pp. 397–416, Columbia Univ. Press, New York.

Francis, J. E., A. Ashworth, D. J. Cantrill, J. A. Crame, J. Howe, R. Stephens, A.-M. Tosolini, and V. Thorn (2008), 100 million years of Antarctic climate evolution: Evidence from fossil plants, in *Antarctica: A Keystone in a Changing World. Proceedings of the 10th International Syposium on Antarctic Earth Sciences*, edited by A. K. Cooper et al., pp. 19–27, Natl. Acad. Press, Washington, D. C.

Francis, J. E., et al. (2009), From greenhouse to icehouse—The Eocene/Oligocene in Antarctica, in *Antarctic Climate Evolution*, edited by F. Florindo and M. Siegert, *Dev. Earth Environ. Sci.*, *8*, 309–368.

Hall, S. A. (1977), Cretaceous and Tertiary dinoflagellates from Seymour Island, Antarctica, *Nature*, *267*, 239–241.

Hannah, M. J., G. S. Wilson, and J. H. Wrenn (2000), Oligocene and Miocene marine palynomorphs from CRP-2/2A drillhole, Victoria Land Basin, Antarctica, *Terra Antart.*, *7*, 503–511.

Hannah, M. J., F. Florindo, D. M. Harwood, C. R. Fielding, and Cape Roberts Science Team (2001a), Chronostratigraphy of the CRP-3 drillhole, Victoria Land Basin, Antarctica, *Terra Antart.*, *8*, 615–620.

Hannah, M. J., J. H. Wrenn, and G. J. Wilson (2001b), Preliminary report on early Oligocene and Latest Eocene marine palynomorphs from CRP-3 drillhole, Victoria Land Basin, Antarctica, *Terra Antart.*, *8*, 383–388.

Ivany, L. C., S. Van Simaeys, E. W. Domack, and S. D. Samson (2006), Evidence for an earliest Oligocene ice sheet on the Antarctic Peninsula, *Geology*, *34*(5), 377–380.

Kennett, J. P. (1977), Cenozoic evolution of Antarctic glaciation, the circum-Antarctic ocean, and their impact on global paleoceanography, *J. Geophys. Res.*, *82*, 3843–3860.

Liu, Z., M. Pagani, D. Zinniker, R. DeConto, M. Huber, H. Brinkhuis, S. R. Shah, R. M. Leckie, and A. Pearson (2009), Global cooling during the Eocene-Oligocene climate transition, *Science*, *323*, 1187–1190, doi:10.1126/science.1166368.

Mao, S., and B. A. R. Mohr (1995), Middle Eocene dinocysts from Bruce Bank (Scotia Sea, Antarctica) and their paleoenvironmental and paleogeographic implications, *Rev. Palaeobot. Palynol.*, *86*, 235–263.

Raine, J. I. (1998), Terrestrial palynomorphs from Cape Roberts Project drillhole CRP-1, Ross Sea, Antarctica, *Terra Antart.*, *5*(3), 539–548.

Raine, J. I., and R. A. Askin (2001), Terrestrial palynology of Cape Roberts Project drillhole CRP-3, Victoria Land Basin, Antarctica, *Terra Antart.*, *8*(4), 389–400.

Sluijs, A., H. Brinkhuis, C. E. Stickley, J. Warnaar, G. L. Williams, and M. Fuller (2003), Dinoflagellate cysts from the Eocene/Oligocene transition in the Southern Ocean: Results from ODP Leg 189, [online], *Proc. Ocean Drill. Program Sci. Results*, *189*, 42 pp. [Available at http://www-odp.tamu.edu/publications/189_SR/104/104.htm.]

Smith, R. T., and J. B. Anderson (2011), Seismic stratigraphy of the Joinville Plateau: Implications for regional climate evolution, in *Tectonic, Climatic, and Cryospheric Evolution of the Antarctic Peninsula*, doi:10.1029/2010SP000980, this volume.

Thorn, V. C., J. B. Riding, and J. E. Francis (2009), The Late Cretaceous dinoflagellate cyst *Manumiella*— Biostratigraphy, systematics, and palaeoecological signals in Antarctica, *Rev. Palaeobot. Palynol.*, *156*(3–4), 436–448.

Truswell, E. M., and M. K. Macphail (2009), Polar forests on the edge of extinction: What does the fossil spore and pollen evidence from East Antarctica say?, *Aust. Syst. Bot.*, *22*(2), 57–106.

Warny, S., J. H. Wrenn, P. J. Bart, and R. A. Askin (2006), Palynological analysis of the NBP03-01A transect in Northern Basin, western Ross Sea, Antarctica: A late Pliocene record, *Palynology*, *30*, 151–182.

Warny, S., R. Askin, M. Hannah, B. Mohr, I. Raine, D. M. Harwood, F. Florindo, and the SMS Science Team (2009), Palynomorphs from a sediment core reveal a sudden remarkably warm Antarctica during the middle Miocene, *Geology*, *37*(10), 955–958, doi:10.1130/G30139A.1.

Williams, G. L., H. Brinkhuis, M. A. Pearce, R. A. Fensome, and J. W. Weegink (2003), Southern Ocean and global dinoflagellate cyst events compared: Index events for the Late Cretaceous-Neogene, [online], *Proc. Ocean Drill. Program Sci. Results*, *189*, 98 pp. [Available at http://www-odp.tamu.edu/publications/189_SR/107/107.htm.]

Wrenn, J. H., and S. W. Beckmann (1982), Maceral, total organic carbon, and palynological analyses of Ross Ice Shelf project Site J9 cores, *Science*, *216*, 187–189.

Wrenn, J. H., and G. F. Hart (1988), Paleogene dinoflagellate cyst biostratigraphy of Seymour Island, *Mem. Geol. Soc. Am.*, *169*, 321–448.

Zachos, J., M. Pagani, L. Sloan, E. Thomas, and K. Billups (2001), Trends, rhythms, and aberrations in global climate 65 Ma to present, *Science*, *292*, 686–693.

Zachos, J. C., G. R. Dickens, and R. E. Zeebe (2008), An early Cenozoic perspective on greenhouse warming and carbon-cycle dynamics, *Nature*, *451*, 279–283, doi:10.1038/nature06588.

R. Askin, 1930 Bunkhouse Dr., Jackson, WY 83001, USA.

S. Warny, Department of Geology and Geophysics and Museum of Natural Science, Louisiana State University, Baton Rouge, LA 70803, USA. (swarny@lsu.edu)

AGU Category Index

Index

Note: Page numbers with italicized *f* and *t* refer to figures and tables